AF812216

Physical Chemistry in Action

Physical Chemistry in Action presents volumes which outline essential physicochemical principles and techniques needed for areas of interdisciplinary research. The scope and coverage includes all areas of research permeated by physical chemistry: organic and inorganic chemistry; biophysics, biochemistry and the life sciences; the pharmaceutical sciences; crystallography; materials sciences; and many more. This series is aimed at students, researchers and academics who require a fundamental knowledge of physical chemistry for working in their particular research field. The series publishes edited volumes, authored monographs and textbooks, and encourages contributions from field experts working in all of the various disciplines.

Miloslav Pekař

The Essentials of Thermodynamics

Springer

Miloslav Pekař
Faculty of Chemistry
Institute of Physical and Applied Chemistry
Brno University of Technology
Brno, Czech Republic

ISSN 2197-4349 ISSN 2197-4357 (electronic)
Physical Chemistry in Action
ISBN 978-3-031-60320-4 ISBN 978-3-031-60321-1 (eBook)
https://doi.org/10.1007/978-3-031-60321-1

© The Editor(s) (if applicable) and The Author(s), under exclusive license to Springer Nature Switzerland AG 2024

This work is subject to copyright. All rights are solely and exclusively licensed by the Publisher, whether the whole or part of the material is concerned, specifically the rights of translation, reprinting, reuse of illustrations, recitation, broadcasting, reproduction on microfilms or in any other physical way, and transmission or information storage and retrieval, electronic adaptation, computer software, or by similar or dissimilar methodology now known or hereafter developed.
The use of general descriptive names, registered names, trademarks, service marks, etc. in this publication does not imply, even in the absence of a specific statement, that such names are exempt from the relevant protective laws and regulations and therefore free for general use.
The publisher, the authors and the editors are safe to assume that the advice and information in this book are believed to be true and accurate at the date of publication. Neither the publisher nor the authors or the editors give a warranty, expressed or implied, with respect to the material contained herein or for any errors or omissions that may have been made. The publisher remains neutral with regard to jurisdictional claims in published maps and institutional affiliations.

This Springer imprint is published by the registered company Springer Nature Switzerland AG
The registered company address is: Gewerbestrasse 11, 6330 Cham, Switzerland

If disposing of this product, please recycle the paper.

Essentially, all known physical laws
were discovered by watching
how things move,
whether they are planets or atoms[1]

[1] Michor F., Liphardt J., Ferrari M., Widom J. What does physics have to do with cancer? *Nature Reviews* **11**, 657–670 (2011).

Dedicated to the memory of Ivan Samohýl
(www.fch.vut.cz/vav/sapelite/is)

Preface

Why another book on thermodynamics? Well, because each author is convinced that he will finally interpret it better, clearer, more comprehensibly than anyone before. This text has at least one more motivating source. It is meant to serve as a precursor to our not-so-long-ago published, more advanced monograph on the non-equilibrium thermodynamics of linear fluids [1]. A precursor, which, similarly to the monograph, draws a great deal from C. A. Truesdell's[2] efforts to cleanse thermodynamics from strange, too intuitive, even mystical procedures and to place it on a solid mathematical foundation such as mechanics. This foundation, however, is also firmly connected with practical knowledge and experience, which it simply tries to generalize or to model in a correct mathematical way.

Despite the attempt to follow a correct mathematical procedure, perhaps not perfect in this basic implementation, this writing attempts to be clear, illustrative and maintains a connection with the experience of both common and experimental reality. Therefore, it resorts quite often to a historicizing approach, so that readers do not get the impression that equations, relations and quantities simply appear in the mind of thoughtful scientists; on the contrary, they will get a basic idea of the building of thermodynamics over time and its logical development. This should contribute to the aforementioned better clarity and understanding of what thermodynamics is all about. As W. F. Giauque[3] said—thermodynamics is not difficult if you can keep track on what it is you are talking about. Historical considerations in the interpretation are recommended by educational experts in chemistry [2].

And yet another insight of contemporary pedagogy readers will find applied here. The interpretation is not strictly a sequence of logically closed, single-topic units, but a single stream of reasoning in which related topics may occur in different places, as the logic of the mainstream requires. This instructional approach appears to promote

[2] Clifford Ambrose Truesdell (1919–2000), American mathematician, creator and promoter of rational thermodynamics, spent most of his academic career at Johns Hopkins University.

[3] William Francis Giauque (1895–1982), American physical chemist, was a professor at the University of California, Berkeley, best known for his research on very low temperatures, for which he received the 1949 Nobel Prize in Chemistry.

desirable rehearsal, leading to a more permanent retention of the discussed content in memory [3–5]. The progression of explanation through completed and closed topics may show better scores on immediate tests of learning, but it affects long-term memory less effectively. Ideally, this book is primarily concerned with permanently embedding the essence of thermodynamics in readers' memories.

One last motivating element. Although the substantial part of the interpretation concerns classical thermodynamics, we systematically include time as the fundamental quantity. This is in contrast to the traditional interpretation where time has completely disappeared, and in line with, for example, Truesdell's approach or Müller's position [6]: "Traditionally, starting with Clausius, there is the strangest reluctance to mention time in thermodynamics, although there is no good reason for the omission of time."

In (text)books like this, it is not customary to place references at every point, but to give a summary of the sources of inspiration. This is also the case here; direct references in the text are limited to necessary cases; most items listed inspired the entire text. Two principal sources used throughout this text were [7, 8], more advanced mathematics was taken from [9–12], personal lockets from [13–16].

There is an accompanying website to this book, please look at https://www.fch.vut.cz/en/rad/poester.

Brno, Czech Republic Miloslav Pekař

References

1. Pekař, M., Samohýl, I.: The thermodynamics of linear fluids and fluid mixtures. Springer, Cham (2014). See also https://www.fch.vut.cz/vav/sapelite
2. Olsson, K.A., Balgopal, M.M., Levinger, N.E.: How did we get here? Teaching chemistry with a historical perspective. J. Chem. Edu. **92**(11), 1773–1776 (2015)
3. Dunlosky, J., Rawson, K.A., Marsh, E.J., Nathan, M.J., Willingham, D.T.: Improving students' learning with effective learning techniques: promising directions from cognitive and educational psychology. Psycholog. Sci. Pub. Int. **14**(1), 4–58 (2013)
4. Karpicke, J.D., Blunt, J.R.: Retrieval practice produces more learning than elaborative studying with concept mapping. Science **331**, 772–775 (2011)
5. Roediger, H.L., Pyc, M.A.: Inexpensive techniques to improve education: applying cognitive psychology to enhance educational practice. J. App. Res. Mem. Cogn. **1**(4), 242–248 (2012)
6. Müller, I.: Max Planck-a life for thermodynamics. Ann. Phy. **17**(2–3), 73–87 (2008)
7. Truesdell, C.: The tragicomical history of thermodynamics 1822–1854. Springer, New York (1980)
8. Truesdell, C., Bharatha, S.: The concepts and logic of classical thermodynamics as a theory of heat engines, rigorously constructed upon the foundation laid by S.Carnot and F. Reech. Springer, New York (1977)
9. Jarník, V.: Differential Calculus I. Praha: Academia (1974). In Czech
10. Kvasnica, J.: Mathematics in physics. Praha: Academia (1989). In Czech
11. McQuarrie, D.A.: Mathematical methods for scientists and engineers. Sausalito: University Science Books (2003)
12. Rektorys, K. et al.: Overview of applied mathematics. Praha: SNTL (1981). In Czech
13. Laidler, K.J.: The world of physical chemistry. Oxford: Oxford University Press (1995)

14. Partington, J.R.: A short history of chemistry. New York: Dover (1989)
15. Pitzer, K.S., Shirley, D.A.: William Francis Giauque 1895–1982. A biographical memoir. Washington: National Academies Press (1996)
16. Tanner, R.I., Walters, K.: Rheology: An historical perspective. Amsterdam: Elsevier, (1998)

Contents

Chapter 1
Introduction

In the summer of 1827, a Scottish botanist, Robert Brown,[1] observed an interesting phenomenon in a microscope, which he also published about a year later [1]. Brown watched the pollen grains scattered in a drop of water and noticed their chaotic, jerky, incessant motion. It was also the time of the rebirth of the atomic theory of the structure of matter. About 80 years later, Einstein[2] and Smoluchowski[3] explained this phenomenon as the result of the permanent motion of water molecules (composed of oxygen and hydrogen atoms). Vast quantities of these building blocks of water are constantly bumping into tiny pollen grains. An enormous number of chaotic impacts move the grain back and forth, and it travels through the liquid medium on a winding, shaky and constantly wobbling path. Brown was probably not the first to observe such *motion* in a microscope, Einstein or Smoluchowski were not the first to explain the behavior of material bodies or solids by means of the unobservable, tiny particles that make up those bodies, but Brown's motion is historically the best-known case of observing microscopic motion (or its consequences) with the naked eye (albeit supported by lens). Microscopic motion has thus taken a firm place in our understanding and description of the events and phenomena we observe around us, alongside macroscopic motion, that is, the motion of bodies actually visible to the naked eye. By microscopic motion we mean the motion of the structural particles of material bodies, not the motion associated with the microscope as an instrument. In the case of water, these particles are molecules, and sometimes we will loosely confuse the terms particles and molecules. After all, the etymology of the word molecule lies in the expression for very small particle, very small matter in Latin or French.

[1] (1773–1858), one of the first to describe the nucleus of plant cells.

[2] Albert Einstein (1879–1955), a famous German physicist, was a professor at Princeton University in the USA from 1933; he is known as the creator of the theory of relativity but was awarded the Nobel Prize in Physics in 1921 for his discovery of the photoelectric effect.

[3] Marian Smoluchowski (1872–1917), an Austrian-born Polish physicist and one of the pioneers of statistical physics, worked at the universities of Lvov and Cracow.

© The Author(s), under exclusive license to Springer Nature Switzerland AG 2024
M. Pekař, *The Essentials of Thermodynamics*, Physical Chemistry in Action,
https://doi.org/10.1007/978-3-031-60321-1_1

Brownian motion—of pollen or similar particles—is a consequence of motion—of molecules—which we have come to call *thermal motion* (and which we certainly cannot see with a light microscope alone). The adjective thermal refers to the commonly used concept of heat, which, as we shall soon see, also resonates in the term of the science we want to discuss here—thermodynamics. Furthermore, it probably also wants to capture an experimentally observed fact—thermal, or Brownian, motion responds to changes in temperature, and is more intense (faster) at elevated temperatures. Temperature (and its changes) is a concept that we associate closely with heat.

When the water molecules hit the pollen grains they "transfer part of their motion" to the pollen grains, thus moving them or changing the direction or speed of their motion. If we already know something of mechanics, let's say, somewhat more technically, that the particles (molecules) impart to the grains some of their energy, which is a certain measure of motion.

Temperature, heat and energy are central concepts of thermodynamics. Why? The translation of the name "thermodynamics" itself suggests this. It is formed from the Greek words heat (*θερμός*) and force (*δύναμις*). So thermodynamics could be the study of the force of heat, which indeed corresponds to the time of its formation, its origin and that coincidentally is also roughly the first half of the 19th century. That was the time when steam engines, devices that really harnessed the "force of heat", were developing rapidly. Nowadays, thermodynamics has progressed far beyond the steam engine, as we shall show in turn. But first, let's stop with those central concepts and look at their meaning and significance. Before doing so, let us just say that thermodynamics can be understood simply as a description of the thermal behaviour of bodies, of the consequences of thermal motion, as a science that handles all our experiences of heat and cold.

Reference

1. Duplantier, B.: Brownian motion, "diverse and undulating". Progr. Math. Phys. **47**, 201–293 (2006)

Chapter 2
Basic Concepts. Thermal Experience

Abstract Basic concepts of human experience with hotness or coldness are introduced—temperature and heat. They are characterized empirically and approached operationally. The thermometer and its role are also discussed. The differences between temperature and heat, two sides of the same coin, are stressed.

2.1 The Way to Temperature and Thermometer

Each of us has personal experience of feeling hot or cold. We feel that warm warms, hot burns, cold cools and frosty really freezes (and as a result, it can also "burn"). These subjective feelings are not always common or distinguishable—in the same room, in the same place, one person may feel cold, another warm and a third "just right". However, in many cases, we can agree on a ranking of bodies from coldest to warmest. Everyone will agree that water in a summer pool is warmer than water frozen in ice, but colder than boiling water. We can no longer distinguish minor differences—e.g., similarly sorting differently cold ice or differently lukewarm water. But it is only a question of the imperfection of our senses. If we can find a more sensitive and objective (feeling-independent) measure, we can rank the bodies from coldest to warmest, preferably by means of a suitable numerical quantity which this measure will give us:

$$\textit{body 1 warmer than body 2} \quad \Leftrightarrow \\ \textit{numerical quantity of the body 1} > \textit{numerical quantity of body} \qquad (2.1)$$

This numerical quantity is called *temperature*. If we denote it in this section with a letter θ, we can rewrite relation (2.1) in a more concise and illustrative way:

$$\textit{body 1 warmer than body 2} \quad \Leftrightarrow \quad \theta_1 > \theta_2 \qquad (2.2)$$

(θ_i is the temperature of body i) or generalize it to some extent:

© The Author(s), under exclusive license to Springer Nature Switzerland AG 2024
M. Pekař, *The Essentials of Thermodynamics*, Physical Chemistry in Action,
https://doi.org/10.1007/978-3-031-60321-1_2

Fig. 2.1 An antique toy; the germ of a thermometer

$$\textit{body 1 is not colder than body 2} \quad \Leftrightarrow \quad \theta_1 \geq \theta_2 \tag{2.3}$$

We know from everyday life that this measure has been found (and not only in one version) and "translated" into a device called a *thermometer*.[1] Let's briefly mention what made it happen. Our ancestors noticed not only their own feelings of heat and cold but also that they were reflected in the properties of the bodies of our world and their changes. Already in antiquity it was supposedly a popular toy, which serves us here as a nice example (Fig. 2.1). It consisted of two vessels, amphorae, connected by a bent pipe passing through their necks. In one of the amphorae the pipe was nicely sealed around the neck, in both of them its ends reached below the water level, which partially filled both vessels. After this, multi-part body had been transferred from the cellar, where one felt the cold, to the sun, where one felt the heat, the ancient onlookers observed a small miracle: the water from the sealed amphora began to rise up by itself through the connecting pipe and move into the other vessel. After returning back to the cellar, it would go back "uphill" on its own again.

The explanation was and is simple based on changes in gas volume as it heated or cooled. The heat of the sun expands the air in the vessel and its increasing volume pushes water into the pipe. As the cellar cools, the volume of air decreases and the water is sucked back into the released volume of the vessel by vacuum.

This toy is actually the germ of the first thermometer (which, however, did not appear until many centuries later). Its essence was to measure temperature based on the volume of gas under given warm-cold conditions.

Other properties that respond to heating or cooling—e.g., electrical resistance and spectral characteristics—can also be used to measure temperature. What thermometers have in common is that they do not actually measure temperature directly, but a suitable property to which we assign a suitable numerical value—the corresponding temperature value.

[1] Try to write a formulation for body 1, which is not hotter than body 2 and deduce the formulation for equally hot bodies 1 and 2 from both versions.

2.2 Temperature and Thermometer in Practice

We will not go into a more extensive (theoretical) discussion of the introduction or definition of temperature here, we will just add a few more practical remarks and refer to the previously published book [1] for a slightly more detailed discussion.

First of all, we will be satisfied that we can measure temperature with a suitable thermometer, and we will regard as temperature what the thermometer measures. Suitable means that it is fast enough (for the processes whose temperature we want to determine) and small or large enough (relative to the body we want to measure). In its—at least theoretical—construction, thermal equilibrium had to be used, the essence of which we will illustrate with a simple example. If we want to determine the value of the property measured by the thermometer corresponding to the melting temperature of water (0 °C) and to mark this value with the corresponding temperature value, we have to immerse the thermometer in melting ice and wait for its "steady state", i.e., wait until its temperature equals the melting temperature of water (more precisely, until the property by means of which we measure the temperature does not change). If we measure the temperature, for example, by the height of the column of liquid in the capillary, we wait until the height is constant, then we can make a mark on this height corresponding to 0 °C; in a similar way, we calibrate the thermometer—provide it with a scale—by means of another point(s) of reference. Thus, thermal equilibrium occurs between bodies in contact that no longer heat or cool each other and thus have the same temperature.

However, we will not limit the temperature we introduced in this way to the equilibrium, our suitable thermometer will serve also for events taking place in time. Let us illustrate this with two examples.

First example. Supposing we want to measure the temperature of water well stirred in a pot while it is being heated to boiling on the cooker for half an hour. We have a dimensionally appropriate thermometer, immersed in that water, which we have found to "react within a microsecond". That is, if we move it from an environment of some constant temperature (which it stably indicates) to an environment of another constant temperature, the temperature value indicated by it will change to the new (and correct) value within a microsecond. What such a thermometer shows you during the above heating process, and when reading, say, after every second, we will safely consider as time-varying values of the temperature of the heated water.

Second example. Prepare two water baths and a similar thermometer. One bath will be at room temperature, the other will contain boiling water. We dip the thermometer into the first bath and after it has settled it reads, say, 21 °C. Move it to the second bath and observe the temperature reading rise until it reaches 100 °C, at which point the thermometer settles. We shall not regard the variable temperature readings taken by the thermometer during heating as absurd, undefined nonsense, but as real values at that moment. However, we will obviously not consider them as temperatures belonging to the water in the bath, but as actual temperatures belonging to the thermometer being heated. If the data from the thermometer is collected by a computer, for example, and every millisecond, the resulting time dependence of temperature will be a realistic curve showing the change in temperature of the thermometer. We can support this reasoning with a complementary thought experiment. Let's prepare another bath with a stable temperature somewhere between the temperatures of the two baths already mentioned—for example 72 °C. When the thermometer immersed in boiling water reaches 72 °C, move it very (infinitely) quickly (without changing it) to the third bath—the reading will certainly remain at 72 °C.

Let's summarize the temperature issue. Temperature is a quantity that describes the hotness and the coldness of bodies, it allows us to rank them from the coldest to the hottest. It does not depend on the size of the body, it cannot simply be added (e.g., if you mix two equal volumes of liquid at different temperatures, the resulting temperature is not simply the sum of the initial temperatures). Usually, we will work with the so-called absolute temperature, measured in kelvin, for which the following applies $T \geq 0$ K.

2.3 Temperature Is Not Enough; Heat

However, temperature alone is not enough to affect a person's experience of hotness or coldness. Anyone who has ever baked knows that we can stick a bare hand in the hot air of a hot oven for a while and it does not hurt us at all. But we cannot touch a baking dish in the oven with a bare hand because it will burn immediately. The baking dish is at the same temperature as the air. Same temperature, but a different experience of hotness.

If we heat water in a pot on the cooker while measuring its temperature, we see it gradually rise to the boiling point (100 °C); there its rise stops until the water boils away. Nothing has changed with the cooker, it still heats the same. And another similar example. If we prepare a mixture of water and ice at 0 °C in a pot, put it back on the cooker and stir it diligently, the temperature will not change until the ice melts—only then will it rise as in the previous example. Again—always the same heating, different "temperature responses".

In everyday life, we say that the cooker transfers "something" to the pot and its contents, which we call "*heat*". Although this is a very familiar view, it is not a very happy one, for it gives the impression that something, a piece of matter or substance perhaps, is really passing from the body of higher temperature to the body of lower temperature. Later on, we shall specify this matter, but for the time being let us be satisfied with the fact that behind the heating or cooling associated with the rise or fall of temperature, or with changes such as melting or boiling, solidification or condensation, there is another quantity called heat. The apparent something that bodies of unequal temperature mutually exchange. Heat, unlike temperature, depends on size and can be added. It takes "more heat" to heat a large pot of water than to heat a small pot by the same temperature difference [2]. You need "twice as much heat" to boil two litres of water than one litre. The gas or electricity bill for the amount of gas or electricity used in heating the cooker can be the measure of heat here.

HEAT	TEMPERATURE
• cannot be assigned to a body in a given state	• can be assigned to a body
• is assigned to a process (heating, cooling)	• does not depend on size (quantity)
• depends on size (quantity)	• is not additive
• is additive	

Fig. 2.2 Comparison of heat and temperature as quantities describing our experience of hotness or coldness

Heat and temperature are two sides of the same coin. The coin we use to describe and quantify our experience of heat and cold, the feelings that accompany them. Hot air imparts "little heat" to the hand compared to an equally hot baking pan; "heat is consumed" first in melting the ice and only then in heating the resulting water. A comparison of the basic properties of temperature and heat as quantities is given in Fig. 2.2.

References

1. Pekař, M., Samohýl, I.: The Thermodynamics of Linear Fluids and Fluid Mixtures. Springer, Cham (2014). See also https://www.fch.vut.cz/vav/sapelite
2. Armstrong, G.T.: The Calorimeter and its influence on the development of chemistry. J. Chem. Educ. **41**(6), 297–307 (1964)

Chapter 3
Motion and Energy

Abstract Energy as a measure of motion is presented and discussed in its historical context. The two basic forms—kinetic and potential energy—and their mutual conversion are also recapitulated. Thermal motion is also mentioned as the related energy transfer between bodies. The sign convention for energy transfer into or out of a body is given.

3.1 Motion, Force and Their Characterisation

We mentioned the concept of energy at the beginning of the introduction when we referred to thermal motion. Before we return to it in more detail, let us stop at the very concept of motion, which energy is closely related to. Motion means moving from place to place, and its cause, as postulated by Newton,[1] is *force*.[2] Newton did not bother to define force, to explain what force is. He "only" showed the relations between the acting force and the characteristics of the corresponding motion, the laws governing the forces, and described how to treat this concept and quantity. In the form of equations, his "laws" then make it possible to describe and predict motion. After all, the term force was probably chosen from the common language to denote an action which produces motion or deformation, which is the result, so to speak, of various motions of different parts of a body. Kicking a ball sets it in motion from rest (and the force of gravity and friction soon sets it to rest again), and in the moment of kicking this deforms its nicely round shape. So, Newton introduced force as the stuff (the cause) that changes the state of motion or shape of a body. However, practical and scientific use demanded the force to measure, to characterize it by numerical value. Or the motion that results from it.

---

[1] Isaac Newton (1642–1727) world-famous English physicist and mathematician, in particular, founder of classical mechanics. He spent much of his professional life at the University of Cambridge and was appointed Master of the Royal Mint in 1699, for which he resigned his Cambridge post in 1701.

[2] More precisely, force is the cause of a change of motion, or the bringing of a body out of either rest or rectilinear motion with uniform speed.

© The Author(s), under exclusive license to Springer Nature Switzerland AG 2024
M. Pekař, *The Essentials of Thermodynamics*, Physical Chemistry in Action,
https://doi.org/10.1007/978-3-031-60321-1_3

Indeed, a long-running debate or dispute soon developed over the true, correct measure of force or motion [1–5]. We will not describe it here but only sketch the essence of the two points of view in a very simplified way.

What is the proper measure of motion, or force? If force is the cause of motion, then some effect of motion is a natural measure of force. Galileo[3] found that different bodies fall (in a vacuum) in free fall at the same rate, but everyone knows that the effects of their impact (on a bare skull, for example) can vary dramatically, especially according to their mass. Therefore, velocity will not be an appropriate measure of motion, and hence force. But neither will mass alone—it makes a difference whether the same body falls on your thumb from 1 or 10 m. However, some appropriate combination of the speed of motion and the mass of what is moving might already be an acceptable measure. Intuitively, we feel this from the fact that, for example, the destructive effects of motion (the impact of a moving body) are greater the faster the body is moving and the more massive it is.

3.2 The Moot Origins of the "Motion Measurement"

Around the time of Newton, scientists investigated various simple mechanical motions—in addition to free fall, for example, the swinging of pendulums, rolling and impact of spheres; they measured mainly the velocities of the bodies involved (in addition to their mass) [5]. An interesting thing was found during this study—certain quantities remained unchanged, i.e., their total numerical value was preserved. Unfortunately, the experimental results were not unambiguous in this respect, which was the cause of the long-standing dispute about what the true measure of motion (of the forces associated with it) is and what the true quantity that is conserved is. What quantity represents the ability of a moving body to set another, initially stationary body in motion by impact? One of the persons who introduced the cause of this dispute was René Descartes,[4] the other was Gottfried Wilhelm Leibniz[5] [1–5]. According to Descartes (1644), the proper measure of motion is the product of mass and velocity, in today's words momentum[6] (Descartes called this product the quantity of motion). Leibniz (since 1686) opposed that the measure is the so-called *vis viva* (living force), i.e., the product of mass and the square of velocity: mv^2. Descartes based his claim

[3] Galileo Galilei (1564–1642), Italian natural scientist, founder of scientific mechanics, "father of modern physics".

[4] (1596–1650), French philosopher, mathematician and physicist, called the father of modern philosophy, one of the founders of modern science.

[5] (1646–1716) German philosopher, mathematician, polymath, one of the discoverers of differential and integral calculus.

[6] Descartes was not the originator of this idea. It appeared, for example, already in the fourteenth century with Jean Buridan (1300–1358, a French philosopher, logician and physicist, who worked at the University of Paris—Sorbonne), who considered the true measure of motion, not the velocity itself, but the product of the velocity and the quantity of matter, substance (*quantitas materiae*) [6, 7].

on the impacts of bodies and Leibniz had the support of elastic collisions. Descartes knew the conservation of quantity, which he called the quantity of motion, in the impacts of (rigid) bodies, and he considered force as something that allows bodies to change their quantity of motion. Leibniz did not recognize this view and regarded force as something intrinsic to a body, "within a body", that allows it to move or move other bodies. The conservation of the *vis viva* in elastic collisions was discovered empirically by Huygens[7] (as late as) around 1673 (more precisely: he published it because he had been carrying out the relevant experiments since ca. 1656 [8]): the sum of mv^2 of a given system of bodies (e.g., two elastic spheres colliding into each other) remains unchanged regardless of how the system of bodies changes through collisions. Leibniz said that a force should be measured by its effects, i.e., by the height to which it can lift a body of a given mass [4, 5]—from free fall it is known that the trajectory is proportional to the square of the velocity.

The fact that the momentum is a vector quantity,[8] whereas the *vis viva* is a scalar quantity, has also brought confusion into the dispute, but we will not really go into such details, let us just point out that it is the second power of the velocity in the *vis viva* that removes the directional dependence, and was thus for Leibniz a more perfect and general measure.

Let us add that one of the participants in that controversy, Johann Bernoulli,[9] used the word energy in 1717 (in a letter to the French mathematician Pierre Varignon (1654–1722), who was the first to formulate Newton's second law in 1700 in the equation we still use today [9]) to denote the product of a force and the (virtual) displacement of the particle on which the force acts. Such a product is the very essence of the quantity called work today, yet Bernoulli is probably the first to use the term energy in a sense related to its present meaning.[10]

Aristotle[11] was probably the first scholar to use the word energy in his works Metaphysics and Physics (350 BC). It was not in the modern sense, but in his discussion of cause and effect, potential and actual. The Greek word *ἐνέϱγεια* here meant something like activity, effort and diligence. Johann Bernoulli may have taken the

[7] Christiaan Huygens (1629–1695), Dutch physicist, mathematician and astronomer, worked for a time in Paris, founder of the wave theory of light. He did not name the quantity mv^2 himself, nor did he particularly emphasize it.

[8] Vector quantities were not yet properly introduced at the time of the dispute, but it was Huygens, perhaps, who found in his experiments that if motion in only one direction is considered, the sum of *mv* is preserved in collisions.

[9] (1667–1748), Swiss mathematician, physicist and physician, spent most of his professional life at the University of Basel, his epitaph states "Archimedes of his century" [8].

[10] The word energy is also found in Galileo's 1638 work on "two new sciences" [10], but, like Aristotle, in the sense of action, for example, in impact, rather than in the sense of an exact physical quantity. That is, in a similar sense as today in colloquial speech, but in the context of characterizing motion and its consequences.

[11] (384–322 BC) one of the most important Greek philosophers, also getting into the natural sciences, and greatly influencing (not only) philosophical thinking.

word from Aristotle, whose work he certainly knew. Then, in 1807, Thomas Young[12] in his printed lectures on "Natural Philosophy and the Art of Mechanics", named *vis viva* by the word *energy*, and is thus usually regarded as the real introducer of the notion of energy in the present sense into the natural sciences. Even if "his" energy is now only kinetic energy (and lacking the one-half factor).

With our concern for historical roots, let us add Partington's point [11] about the first use of the word energy in its modern sense in the Encyclopédie Méthodique.[13] Its second volume, published in 1785, under the heading "Force", contains the phrase (p. 82) [13] "there is an effort or energy in a moving body which is not present in a body at rest". This is probably not the better-known encyclopaedia edited by Denis Diderot between 1751 and 1772, [14, 15] in which Jean d'Alembert[14] also contributed editorially and as author between 1746 and 1758 [16], but its succeeding version edited by Charles-Joseph Panckouck (and then by his son) from 1782. It is d'Alembert who, according to the cipher at the end of that entry and the list of contributors, is the author of that entry, although the volume was published 2 years after his death. Interestingly, a similar entry in Diderot's Encyclopaedia, also compiled by d'Alembert (vol. VII, pp. 110–120), does not mention energy at all, but still uses the term "living force, force vive"; on the other hand, in the entry "Energetiques" (vol. V, p. 651), he attributes this name only to particles in (rapid) motion, such as particles of gunpowder, and concludes that the term is not used otherwise. This encyclopedia reserves the word energy for rhetorical use only (energetic painting, energetic expression).

3.3 Two Perspectives on Motion and Force Measurement

The essence of the above-mentioned controversy was actually the question of whether to measure force through temporal or spatial effects, whether to take it as a temporal or spatial function. Whether to observe how long the motion, the action of the force, lasts or over what distance the force is able to accomplish the motion. It is quite revealing to look at the mathematical reasoning on both sides of the dispute [7]. The followers of Descartes' camp compared two forces, F_1 and F_2, acting for a given time interval t as follows:

$$\frac{F_1}{F_2} = \frac{m_1 a_1}{m_2 a_2} = \frac{m_1 a_1 t}{m_2 a_2 t} = \frac{m_1 v_1}{m_2 v_2}.$$

[12] (1773–1829), a versatile English physician, interested in physics or philology, demonstrated diffraction or interference of light showing its wave character, his name still sounds in Young's (elasticity) modulus.

[13] This information is somewhat casually adopted by Moore [12].

[14] Jean le Rond d'Alembert (1717–1783), French mathematician, physicist and philosopher.

The forces are therefore in the ratio of the corresponding Descartes' quantities of motion (momenta). The proponents of Leibniz's view argued in measuring forces by the path s along which they act, as follows[15]:

$$\frac{F_1}{F_2} = \frac{m_1 a_1}{m_2 a_2} = \frac{m_1 a_1 s}{m_2 a_2 s} = \frac{\frac{1}{2} m_1 v_1^2}{\frac{1}{2} m_2 v_2^2} = \frac{m_1 v_1^2}{m_2 v_2^2}.$$

The forces are then in proportion to the corresponding *vis viva*.

In modern terms, we would express this measurement of motion or associated force by integration. In the "Descartes's" approach, as follows:

$$\int F \mathrm{d}t = \int (m\, \mathrm{d}v/\mathrm{d}t)\mathrm{d}t = mv\ (+const.), \tag{3.1}$$

in the "Leibniz's" approach in this way:

$$\int F \mathrm{d}s = \int (m\, \mathrm{d}v/\mathrm{d}t)\mathrm{d}s = \int (m\, \mathrm{d}s/\mathrm{d}t)\, \mathrm{d}v = \int mv\, \mathrm{d}v = \frac{1}{2} mv^2\ (+const.) \tag{3.2}$$

(in both cases we assumed constant mass). In the first case, we get the momentum as a measure (here simplified only as the magnitude of the vector quantity), in the second case the quantity *vis viva* in today's concept—i.e., with one half as a coefficient and called *kinetic energy*. Historically, however, Coriolis[16] (in 1829 [7]) added the coefficient of one half to the *vis viva* just because of this integration, which relates it to the work done (given by the integral $\int F \mathrm{d}s$); at that time, the measure of work in engineering practice had long been the weight (1000 kg) raised to a height (1 m). But the name kinetic energy instead of *vis viva* was not introduced until Kelvin[17] in 1849 to better distinguish between force and energy.

Both energy and momentum are equally good measures of force or motion, each has its place, and sometimes we cannot do without both to explain observed motions. An example is the so-called Newton's pendulum, shown in Fig. 3.1. Try to prove for yourself, using conservation of momentum and kinetic energy, what happens when one outermost ball is deflected and released and what happens when two balls are deflected and released, and note whether one conservation is sufficient or whether you must use both balls. Momentum and (kinetic) energy are the same in that what one body loses in the impact, the other body gains. That dispute was actually unnecessary, we just have (at least) two measures of motion.

---

[15] It was known from Galileo's experiments that $v = \sqrt{2as}$.

[16] Gustave-Gaspard de Coriolis (1792–1843), French physicist and mathematician, discoverer of the force in rotating bodies that now bears his name.

[17] William Thomson, by his noble name of Lord Kelvin of Largs (1824–1907), a distinguished Scottish physicist, was a professor at the University of Glasgow, involved in electrostatics, magnetism, thermodynamics, hydrodynamics and geophysics, and was interested in practical applications, contributing to a number of patents. The unit of absolute temperature, which he conceived, is named in his honour.

Fig. 3.1 The scheme of the Newton's pendulum

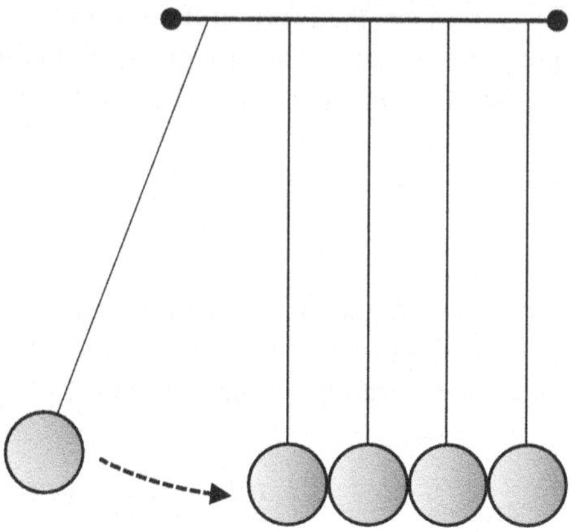

3.4 Energy Is Conserved, It Strengthens Its Place

So in rigid elastic body mechanics, kinetic energy is conserved—it remains constant no matter how the system of bodies changes through collisions. But if there are forces acting at a distance, this is no longer the case. If we throw a body upwards (in the Earth's gravity field), we find that it gradually slows down, stops briefly at its highest point, and then begins to accelerate in the opposite direction. Its kinetic energy therefore decreases, is even zero for a while, and then increases again. In this reverse motion, the body behaves in exactly the same way as if we first held it at its highest point and then released it. It was already claimed by Leibniz (1695) that bodies in motion have *vis viva*, while bodies at rest, which are lifted (in a gravitational field) or stretched or compressed (e.g., a spring), have *potentia* or potential force [6]. This can manifest itself by motion as soon as conditions permit (we release a lifted body or compressed spring). In 1742, Daniel Bernoulli[18] writes in a letter to Euler[19] about the difference between the potential *vis viva* of a stretched elastic cord and the *vis viva* appearing when the cord is released [17]; Euler himself writes about *vis potentialis* in 1744. Similarly, Lazare Carnot[20] called the product of mass and height by the term latent *vis viva*; he argued that a body at a certain height has a certain form of *vis viva*, and that a stretched spring contains a certain amount of *vis viva*, and that this latent part of it can be transformed into actual *vis viva*.

[18] (1700–1782), Swiss mathematician of wide natural scientific interests, son of Johann Bernoulli, founder of hydrodynamics, his main place of activity was the University of Basel.

[19] Leonhard Euler (1707–1783), a Swiss mathematician and physicist, a very prolific pioneer of pure mathematics and its applications in mechanics, worked in St. Petersburg for a long time, where he died and is buried.

[20] Lazare Nicolas Marguerite Carnot (1783–1803), a French politician, general and military engineer, made scientific contributions, especially in the field of geometry.

Julius Mayer[21] (in 1842) and Hermann Helmholtz[22] (in 1847) eventually realized that we must add that latent form of *vis viva* to *vis viva*, and only their sum is generally preserved. In the example of the upward throw, the two quantities are nicely interchanged—just how much one decreases, the other increases. In 1853, William Rankine[23] introduced the name *potential energy*[24] for the latent form, allowing Helmholtz to state that a decrease in potential energy is (for interacting bodies) equal to an increase in kinetic energy, and *vice versa*. And finally, Maxwell[25] in 1877 writes: the kinetic energy is given by motion, the potential energy is given by configuration.

3.5 Energy as a Measure of Motion

Energy is thus one of the quantities describing motion, both actual and potential. It is a quantitative measure of motion for both a moving body and a body at rest placed in a force field. It has no material character, it is not something concrete that exists in itself, it is not a thing or a process, it is a property. It is an abstract quantity having the meaning of a numerical characteristic assigned to a body, it does not exist in itself. It is nothing special, for example, speed or mass do not exist as separate substances either, but they are numerical characteristics of certain properties of a certain body. Energy has different forms [19, 20], so far we have known kinetic and potential energy, these can be transformed and converted into each other (Fig. 3.2), but the numerical value of energy does not change in sum. Energy conversion does not mean that something is actually converted,[26] like liquid water into ice. It just means a mutually balanced change in the numerical values of different forms of energy—as much as the value of one form decreases, the value of the other form increases. Further, energy can be transferred from body to body and exchanged between bodies. Again, this transfer does not mean that a body loses a piece of something or gains a piece of something.

---

[21] Julius Robert von Mayer (1814–1878), a German physician interested in thermal physics and energy transformations.

[22] Hermann Ludwig Ferdinand von Helmholtz (1821–1894), a German physiologist, physician and physicist involved in various fields of science, active especially at the universities of Heidelberg and Berlin.

[23] William John Macquorn Rankine (1820–1872), a Scottish engineer and physicist, among other things, analysed the function of the steam engine, professor at the University of Glasgow.

[24] Interestingly, Rankine was familiar with classical philosophy, Aristotle and his concepts of actuality, or Greek energy, and potentiality, and saw them as distant ancestors of (potential) energy [18].

[25] James Clerk Maxwell (1831–1879), a Scottish physicist of wide interests, best known as the author of the general mathematical description of the electromagnetic field or as one of the developers of the kinetic theory, worked for a time at the University of Aberdeen and King's College, London, but after several years spent at the family residence in south-west Scotland, he joined the University of Cambridge in 1871.

[26] We leave aside the energy-mass relationship.

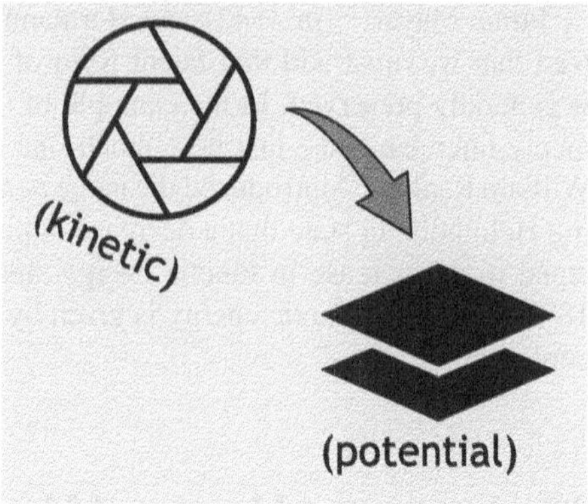

Fig. 3.2 Forms of energy can change from one to another

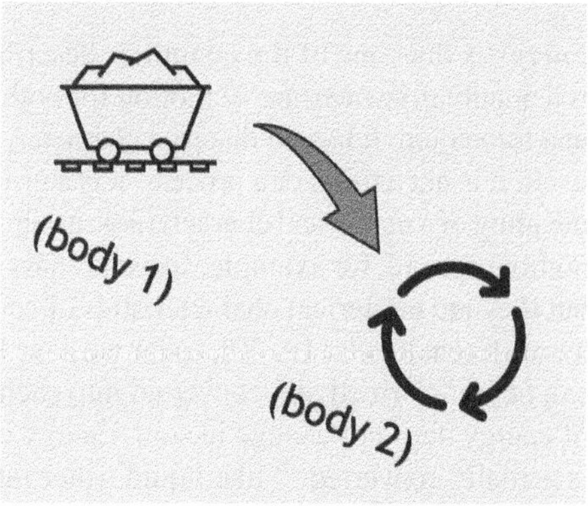

Fig. 3.3 Energy can be transferred between bodies

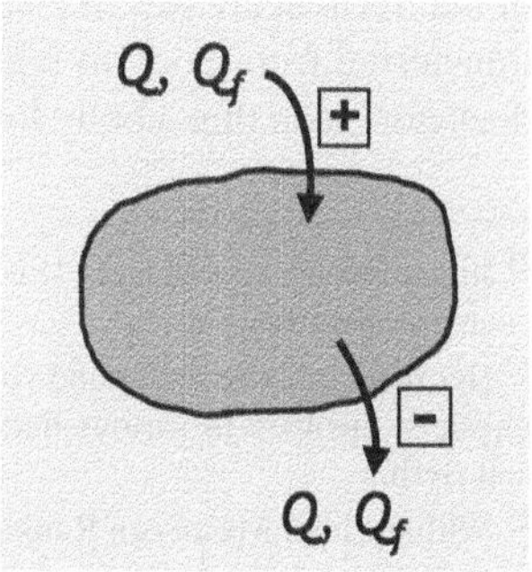

Fig. 3.4 Sign convention for energy transfers illustrated on heat (heating)

Here again, it is just a matter of changing the numerical characteristics—as much as the energy of one body decreases, the energy of the other body increases (Fig. 3.3).

The following convention has been adopted for energy transfer. The transfer to the body under consideration has a positive sign and the transfer from that body has a negative sign. This is illustrated in Fig. 3.4.

3.6 Thermal Motion

Let us return again to thermal motion, more precisely, let us recall the experience related to motion when heated. Anyone who has ever cooked anything will have noticed that as the temperature of, for example, heated water rises, its motion increases—in the right and good lighting, one can see some kind of moving structures, smudges or waves. If there are visible, tiny particles in the water, such as small pieces of herbs or spices, we can also observe their motion and its amplification with increasing temperature. Shortly before boiling, we observe bubbles rising (i.e., moving) near the bottom. The steam rising from the boiling water rolls out of the pot and is able to lift the relatively heavy lid.

We can explain the above motion phenomena using the idea of the molecular composition of matter. As the temperature rises, the molecules move faster and faster, and this can be reflected in the more intense Brownian motion of, for example, the herb particles or in the escape of the already quite fast molecules from the liquid medium into the gaseous state, in which they can also move the lid with their lively motion and impacts.

Similarly, we can look at melting (of ice in cold water, for example, mentioned above), where the temperature does not have to rise as the water-ice body heats up (and it doesn't unless we overheat it). In the solid state, ice, the motion of molecules is obviously restricted; otherwise, solids would not hold their shape. Of course, they can move around a bit without causing a change in shape observable to the eye or microscope. They simply oscillate slightly around their "solid state" positions, in the case of ice, around their positions in its crystal structure.

Melting when heated then means that the molecules are moved in such a way that they are able to leave their "rigid positions", abandon their fixed shape and go into a liquid state. There the molecules move so much that they do not hold a fixed shape, but, as a whole, they conform to the shape of the (solid) container. Even in a liquid, the motion of the molecules is still "miniature", imperceptible to the eye, and the shape given by the container is held. But the fastest molecules can leave the liquid through its surface and enter the gaseous state.

What remains to be thought about is what it is that a hot body (the cooker) does to a colder body that sets its molecules in motion, and so perhaps increases its temperature, volume, or changes its state of matter. We begin our search for the answer by looking at the history of the study of thermal behaviour, which laid the foundations for building thermodynamics as an exact science. Historically, the first, more thoroughly tested experiences of thermal behaviour were gained by people with gaseous bodies, simply, gases.

References

1. Iltis, C.: D'Alembert and the vis viva controversy. Hist. Philos. Sci. **1**(2), 135–144 (1970)
2. Iltis, C.: Leibniz and the vis viva controversy. ISIS **62**(1), 21–35 (1971)
3. Iltis, C.: The decline of Cartesianism in mechanics: the Leibnizian-Cartesian debates. ISIS **64**(3), 356–373 (1973)
4. Iltis, C.: The Leibnizian-Newtonian debates: natural philosophy and social psychology. Br. J. Hist. Sci. **6**(24), 343–377 (1973)
5. Papineau, D.: The vis viva controversy: do meanings matter? Stud. Hist. Philos. Sci. **8**(2), 111–142 (1977)
6. Hecht, E.: An historico-critical account of potential energy: is PE really real? Phys. Teach. **41**(8), 486–492 (2003)
7. Jammer, M.: Concepts of Force. Dover, New York (1999)
8. Lindsay, R.B.: Energy: Historical Development of the Concept. Dowden, Hutchinson & Ross, Stroudsburg (1975)
9. Laidler, K.J.: The World of Physical Chemistry. Oxford University Press, Oxford (1995)
10. Galilei, G.: Dialogues Concerning Two New Sciences. Dover, New York (2014) (reprint of English translation)
11. Partington, J.R.: An Advanced Treatise on Physical Chemistry. The Fundamental Principles - The Properties of Gases, vol. 1. Longmans, London (1949)
12. Moore, W.J.: Physical Chemistry, 4th ed. Englewood Clis: Prentice Hall, 1972. Czech translation, SNTL, Praha (1979)
13. Encyclopédie méthodique, ou par ordre de matieres: par une société de gens de lettres, de savans at d'artistes. https://books.google.fr/books?id=endBAAAAcAAJ&hl=fr&pg=PP7#v=onepage&q&f=false (visited July 6, 2022)
14. Encyclopédie, ou dictionnaire raisonné des sciences, des arts et des métiers, par une société de gens de lettre. http://enccre.academiesciences.fr/encyclopedie/ (visited July 6, 2022)
15. Doig, K.H.: From Encyclopédie to Encyclopédie méthodique revision and expansion. Oxford: Voltaire Foundation, University of Oxford (2013)
16. Hankins, T.L.: Jean d'Alembert. Science and the Enlightenment. Clarendon Press, Oxford (1970)
17. Roche, J.: What is potential energy? Eur. J. Phys. **24**(2), 185–196 (2003)
18. Lewis, C.J.T.: Heat and Thermodynamics: A Historical Perspective. Greenwood Press, Westport (2007)
19. Arons, A.B:. Development of energy concepts in introductory physics courses. Am. J. Phys. **67**(12), 1063–1067 (1999)
20. Bunge, M.: Energy: Between physica and metaphysics. Sci. Ed. **9**(5), 457–461 (2000)

Chapter 4
The Behaviour of Gases

Abstract Pressure as an important variable is discussed first. Historically important experiments on the behaviour of gases under constant temperature or pressure are described and the corresponding equations are given. The latter are then combined into the state equation for an ideal gas. The basic kinetic theory of ideal gases is developed and its expressions for pressure and kinetic energy are presented. The interpretation of temperature in terms of the kinetic energy of gas particles is presented. Heat is explained as a process of energy transfer between bodies. The first attempts to measure heat and to quantify it are described including heat units.

4.1 First, Important Notes on Pressure

Robert Boyle[1] was one of the first to bring more exact knowledge of the (thermal) behaviour of gases in terms of the relationship between their pressure, volume and temperature (at rest, at equilibrium). In the case of gases, pressure is one of the basic quantities we use to describe their behaviour, and it is also a term commonly used in everyday life. However, the exact definition of pressure as a physical quantity is not entirely straightforward. A more detailed discussion can be found, for example, in [1], here we restrict ourselves to the essentials.

Intuitively, by pressure we mean a force that pushes on something. *Pressure* should therefore be another quantity closely related to force. Understandably, it would make no sense to introduce two terms to refer to the same thing (at least not in the exact science we are attempting here). So what "force-specific" do we want to express by pressure? Let's look at a practical example. Standing on or walking on freshly fallen snow in winter, the weight (which is a kind of force) of our body—caused by the Earth's gravitational field, its gravitational force—is exerted on the snow, which does not quite withstand this force, and we sink into the snow. However, if

[1] (1627–1691), an Irish-English natural scientist of wide scope, regarded as the founder of modern chemistry, spent more than 10 years (ca. 1654–1668) in Oxford experimenting with gases, among other things (see Boyle's Law below), then moved to London to his sister's house, where he continued his scientific activities.

© The Author(s), under exclusive license to Springer Nature Switzerland AG 2024
M. Pekař, *The Essentials of Thermodynamics*, Physical Chemistry in Action,
https://doi.org/10.1007/978-3-031-60321-1_4

we put on snowshoes or skis, we sink noticeably less. Same force (our weight; we neglect the weight of skis or snowshoes), but different effect. In the first case, the weight was applied "over" the surface of our soles, and in the second case over the apparently larger surface of the snowshoes or skis. On a larger area, the weight was more "spread out", it also acted on a larger area of snow, so it didn't move as much snow downward and we didn't sink as much. If we divide the (equal) applied gravity by the area, we get the "recalculated" measure of this "distributed" action—pressure. In this concept, the pressure is given by the ratio of the force and the area on which the force acts perpendicularly.

The cause of the (introduction of) pressure in the snow example was thus gravitational force. The situation is similar in liquids. Anyone who has dived a little deeper has felt at first hand—that is, with their ears—the higher pressure at greater depths, caused by the weight of the liquid above that depth. This pressure is called *hydrostatic* because it is typical of a stationary (static) column (layer) of liquid (originally water, hence *hydro*, but the term has been transferred to other liquids). Because liquids do not hold their shape, i.e., they "move over" under the action of some force and are only held in place by sufficiently strong vessel walls, the hydrostatic pressure at a given point in the still liquid is the same in all directions (otherwise it would flow in the direction of the resultant force), and of course, it also pushes on the vessel wall at the corresponding point. This can be ascertained by immersing a suitably inflated balloon (with a radius negligible in relation to the depth of immersion). The pressure will deform it pretty symmetrically, still into a ball (smaller than before immersion) and not into some conical shape. Thus, a round inflated balloon indicates a symmetrical application of pressure to the membrane forming the balloon.

Hydrostatic pressure is quite easy to express in elementary physics, if we realize that the given gravitational force, gravity (its magnitude), is given by the product mg. If the mass of the liquid m acts on an area A, the pressure is given by the fraction mg/A. If the height of the liquid layer is h, we can express the area in terms of its volume: $A = V/h$. After introducing the density of the liquid (ρ), we obtain the resulting expression for the hydrostatic pressure: $\rho g h$.

Hydrostatic pressure was one of the first pressures that our ancestors learned to measure; in doing so, it was closely related to gas pressure, specifically the normal atmospheric pressure of air. Torricelli[2] in 1643 showed that a vertical tube, sealed at one end, filled with mercury and immersed with the open end in an open vessel also filled with mercury, would reach a state in which the height of the mercury in the tube would be stabilized at about 762 mm. Thus, the whole tube does not flow out and a vacuum is "created" between the sealed end and the mercury level. Torricelli explained this by saying that the mercury is held in the tube by the pressure of the surrounding air, thus acting on the mercury level in the open container (and through this liquid, the pressure is "transmitted" to the tube). Boyle called this device (in

[2] Evangelista Torricelli (1608–1647), an Italian physicist and mathematician and the successor of Galileo at the University of Pisa, is known as the author of Torricelli's law and the inventor of the barometer.

1665) a barometer (from the Greek words $\beta\acute{\alpha}\varrho o\varsigma$, weight, $\mu\acute{\epsilon}\tau\varrho o\nu$, measure), which we now regard as an instrument for measuring (atmospheric) pressure.

Boyle also confirmed Torricelli's deductions by using his own improved vacuum pump to extract air from the enclosed space, thus reducing the pressure in that space. He showed that the height of the mercury column of a barometer enclosed in the extracted space decreases as the air pressure decreases. He was able to lower it almost to the level of the barometer's open container.

Simple and probably the first *manometers* (from the Latin *manus*, force, and the Greek $\mu\acute{\epsilon}\tau\varrho o\nu$, measure), i.e., pressure gauges, also worked on the principle of hydrostatic pressure. They consisted of a U-tube filled with a suitable liquid (such as mercury). If a different pressure was applied to the liquid in each arm of the tube, the liquid level would settle at a different height in each arm. From the difference in the heights of the levels, we can calculate the difference in the pressures acting on those arms, knowing the density of the liquid. This is usually enough for us if we know one pressure—for example, the atmospheric pressure (as determined by a barometer).

We could conclude this little pressure trip with a simple statement, a kind of operational definition: pressure is what we measure with a pressure gauge (manometer). We need to know on what principle it measures. In the case of liquids, in the simple case, it is hydrostatic pressure; in the case of gases, it is the pressure detected through hydrostatic pressure—the pressure of the gas pushes the liquid to a certain height. For most of the cases (systems, bodies) that we will deal with here, this will be a quite sufficient view of pressure.

4.2 First Experience with the Thermal Behaviour of Gases

4.2.1 Boyle and Contemporaries

We have already noted that Boyle was one of the pioneers in the investigation of the relation of the pressure of a gas to its volume and temperature. In experiments carried out during 1661 (published a year later [2]), Boyle used a U-shaped (or rather J-shaped [3]) tube. The end of the shorter arm was sealed and he trapped a certain volume of air, which was the studied gas, in it using mercury poured through the longer end. Increasing the height of the column of mercury in the longer end increased the hydrostatic pressure acting on the trapped volume of air. The experiments were carried out at a constant temperature (today we would say in a thermostat). So he also used the hydrostatic pressure of the mercury to determine the gas pressure.

In this way, Boyle subjected air (gas) to pressures higher than atmospheric, but he also worked with lower pressures. To do this, he used a straight glass tube sealed at one end and trapped air in it—above the mercury—for a length of about 2.5 cm. By lifting the tube (closed end up) in the mercury container, he caused the air to expand until it occupied a space of over 80 cm in the tube.

The pressures he was able to achieve in his experiments ranged from (hydrostatic pressure) 30 to 3000 mm of mercury. He found the volume of air from the (internal) cross-section of the tubes and the length of the part of the tube that the air occupied. The results of his measurements led him to conclude that at a given temperature the product of pressure and volume of a given quantity (mass) of air was constant:

$$PV = \text{const.}\ (T = \text{const.}). \tag{4.1}$$

Today, this observation, later extended to other gases, is called *Boyle's law*, although claims to primacy and proper naming have been made by others, and it is sometimes called the Boyle-Mariotte law in the literature. Those interested in a deeper consideration of the matter may read the article [2], of which we give here only the conclusion: the law was discovered by Power[3] and Towneley,[4] was afterwards accurately verified by Hooke[5] and again by Boyle, who was also the first to publish it, and finally most widely publicized by Mariotte.[6] Regnault[7] was then probably the first to generalize experimental experience into the claim that all gases at (sufficiently, very) low pressures satisfy the law (as late as 1847).

Boyle conducted his experiments at a constant temperature. From ancient times (see also Sect. 2.1) another experience was known—gases expand when heated. In other words, they increase in volume as the temperature rises (unless held constant by sufficient pressure). More precise measurements were not made until the late 17th century by Amontons[8] (1699) for heating air in a constant volume. He found that regardless of the initial pressure, the relative change in pressure for a given change in gas temperature (i.e., between two specific, given temperatures) is the same. Boerhaave[9] in 1732 concluded that the expansion of air is proportional to the change in its temperature.

[3] Henry Power (1623–1668), an English physician and experimenter, collaborated with Towneley, a family friend, in his experiments with gases.

[4] Richard Towneley (1629–1668), a member of a prominent Roman Catholic family in Lancashire, England, whose wealth allowed him to pursue fields such as meteorology and astronomy and to collaborate with other experimenters such as Boyle.

[5] Robert Hooke (1635–1703), English naturalist, discoverer of the law of elasticity, which bears his name, or the bending of light. He collaborated with Boyle at Oxford, was a skilled experimenter, and intervened, for example, in microscopy, astronomy, the issues of combustion and respiration.

[6] Edme Mariotte (c. 1620–1684), a French scholar, was a Roman Catholic priest and one of the founders of the French Academy of Sciences. He contributed to the development of experimental science, especially physics and plant physiology.

[7] Henri Victor Regnault (1810–1878), a French chemist and physicist, followed Gay-Lussac in the professorship of chemistry at the École Polytechnique, later serving as director of the famous porcelain factory at Sévres, where he continued to investigate the thermal properties of bodies.

[8] Guillaume Amontons (1163–1705, French engineer, self-taught inventor and instrument designer. He acquired his relationship to and skill in practical mechanics while working on various public—in today's terms—projects.

[9] Hermann Boerhaave (1668–1738), a Dutch physician, botanist and chemist, professor at the University of Leiden, founder of bedside medical education.

4.2.2 *Dalton, Gay-Lussac and Others*

The somewhat peculiar English chemist (and physicist) Cavendish[10] conducted experiments in 1779–80 on the expansion of gases in a Boyle-like U-tube, which he tempered in a water bath. He investigated several gases, including air, nitrogen, oxygen, hydrogen and carbon dioxide. Unfortunately, he did not publish his results (during his lifetime), nor did he reach the generalizable conclusion they pointed to—namely, that the coefficient of expansion per degree of temperature is the same for different gases. The relation describing the expansion of air was probably first published by Volta[11] in 1793.

It was not until 1802 that Dalton[12] published more thorough measurements with dry gases. He used a straight glass tube, sealed at one end, in which a drop of mercury trapped the examined gas. He concluded his measurements by stating that different gases expand to the same extent when heated to the same temperature difference (at constant pressure). At the same time Gay-Lussac[13] carried out similar experiments with a slightly modified apparatus (a glass flask with a long, narrow neck containing a sealing drop of mercury). He reached the same conclusions, which he published about six months later:

$$V_t = V_0(1 + \alpha t) \tag{4.2}$$

Here V_0 is the initial volume (for Gay-Lussac at 0 °C), α is the coefficient of proportionality, t is the temperature in Celsius degrees. According to Partington [4], this law was discovered for oxygen, nitrogen, hydrogen and carbon dioxide as early as about 1787 by Charles,[14] who also did not publish his results but communicated them verbally to Gay-Lussac. Most often this relationship is called *Charles' Law* or *Gay-Lussac Law*. Partington concludes its naming as follows. Amontons and Volta obtained reliable data for air only. Charles, Dalton and Gay-Lussac found virtually the same expansion coefficient for all the gases they used.

The above equation implies that—if valid at all (decreasing) temperatures—at $t = -1/\alpha$ °C the volume V_t is zero. Indeed, the data suggest that extrapolating them for different gases yields a common limit of zero volume at $-273{,}15$ °C.[15] An

[10] Henry Cavendish (1731–1810), a quiet recluse, came from a distinguished aristocratic family, and set up his mansion as a laboratory in which he conducted natural science experiments. He published his results very little, his discoveries bear the names of later "rediscoverers".

[11] Alessandro Giuseppi Antonio Anastasis Volta (1745–1827), an Italian physicist, known for his discoveries in electricity and electrochemical cells, professor at the University of Pavia.

[12] John Dalton (1766–1844), an English teacher and private scholar (chemist, biologist and physicist), active mainly in Manchester, is known as the proponent and creator of the modern atomic theory, and the law of partial pressures of gases bears his name.

[13] Joseph Louis Gay-Lussac (1778–1850), a French chemist of wide interests, professor at the Polytechnic of Paris, (co)discoverer of several compounds or elements.

[14] Jacques Alexandre César Charles (1746–1823) a French physicist, mathematician and inventor, pioneer of ballooning, member of the French Academy of Sciences, professor at Conservatoire des Arts et Métiers.

[15] The first extrapolators sought the value of absolute cold in this way.

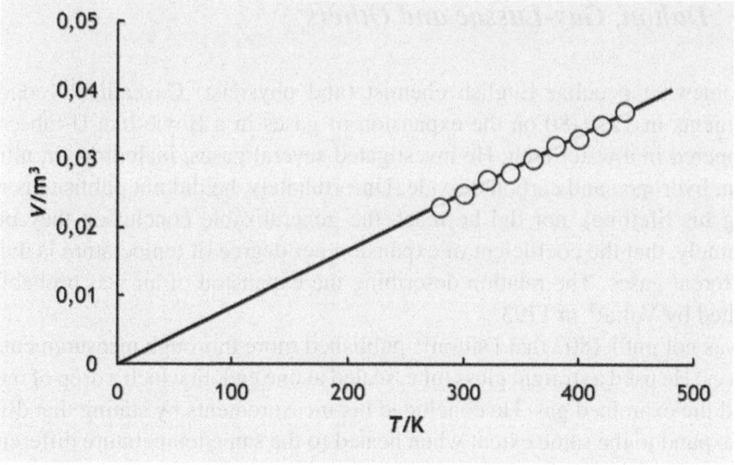

Fig. 4.1 Dependence of the volume of one mole of helium on temperature [5]

example is given in Fig. 4.1 [5]. We will use this observation in a moment after we combine the two experimental laws into a single one that describes the observations of the gases in general. The essence of this connection is indicated by the diagram:

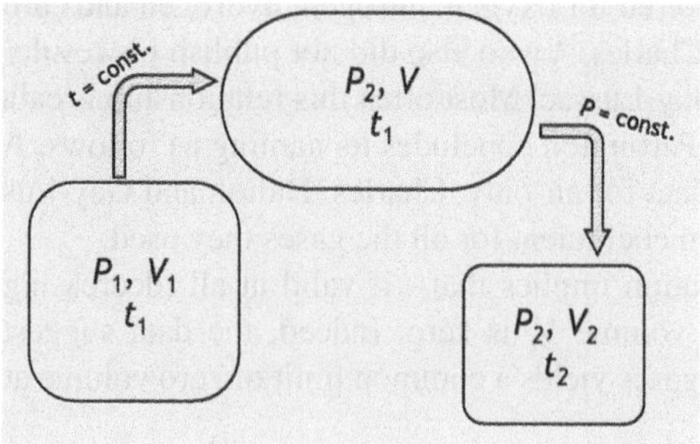

4.2.3 *The Combination of Experiences: The Equation of State*

All operations in the above diagram are performed with a given (constant) amount (i.e., mass) of gas. The first step thus corresponds to Boyle's law

$$P_1 V_1 = P_2 V. \tag{4.3}$$

For the ratio of the volumes occurring in the second step, according to (for example) Charles' law

$$V_2/V = (1 + \alpha t_2)/(1 + \alpha t_1). \tag{4.4}$$

Combining the two relations, we get

$$\begin{aligned} V &= P_1 V_1 / P_2 \\ V_2/V &= V_2 P_2/(P_1 V_1) = (1 + \alpha t_2)/(1 + \alpha t_1) \\ P_2 V_2 &= P_1 V_1 (1 + \alpha t_2)/(1 + \alpha t_1). \end{aligned} \tag{4.5}$$

If we choose (according to Gay-Lussac) $t_1 = 0$ °C and $t_2 = t$ (and adjust the subscripts accordingly), we have $PV = P_0 V_0 (1 + \alpha t) = (P_0 V_0/\alpha)(t + 1/\alpha)$.

Since $P_0 V_0$ has a constant value according to Boyle's law (e.g., for 1 g of gas) and α according to Charles' law, the expression $P_0 V_0/\alpha$ is constant—let us denote it (for 1 g of gas) $\bar{R}$. If we replace the Celsius temperature by the temperature T defined by the above limit: $T = t + 273.15$, we obtain the equation:

$$P V_s = \bar{R} T, \tag{4.6}$$

where we have emphasized the volume of that one gram of gas with the symbol V_s (it is therefore a specific volume). However, the value of the constant $\bar{R}$ also depends on the gas being weighed. The generalization is obtained by considering Avogadro's[16] hypothesis and switching to the molar amount. Avogadro says that equal volumes of different gases contain the same number of particles (molecules) at the same temperature and pressure. In chemistry, we express the number of particles in terms of moles, and the mass of one mole of a gas is expressed by its molar mass M. For one mole of a gas, we then generally have

$$P V_m = M \bar{R} T = RT \tag{4.7}$$

(V_m is the volume of one mole, i.e., the molar volume) and finally for n moles

$$PV = nRT. \tag{4.8}$$

Thus we have reached what is probably the simplest equation describing the behaviour of gases, i.e., the relationship between temperature, volume and pressure (of a given quantity—n) of a gas. It is customary to call such equations *state equations* in classical (equilibrium) thermodynamics, and rather *constitutive equations* in more modern, non-equilibrium thermodynamics [6].

Now let us also point out that we have at the same time introduced a different temperature scale—the absolute temperature scale with the unit kelvin. We introduced it simply via the Celsius scale using extrapolation of the behaviour of gases. This is

[16] Amedeo Avogadro (1776–1856), an Italian physicist, though a lawyer by education, professor of physics at the University of Turin.

something of a roundabout way of doing it, and we should more properly call this the (empirical) temperature of an ideal gas thermometer. Here we will be content with stating that this temperature is the same as the absolute temperature, which is universal and takes positive values and is quite sufficient for our thermodynamic considerations. For those interested in a deeper discussion or even proof, we refer, for example, to the book [6].

4.3 A Simple Model of Microscopic Motion

4.3.1 *Particles in Motion; The Ideal Gas*

Let us leave experimental experience with gases for a moment, return to our introduction, and consider the following—only seemingly remote—question. How could we quite simply model the microscopic (thermal) motion the effects of which Brown observed in water? Imagine a molecule as an abstract material particle of mass m. Let the typical velocity of its motion be v. If it hits something, "by the force of its motion" it can move it, change its state of motion, as Brown observed with pollen grains. Water, however, would be a complex material to simple modelling—their molecules are too many in a given volume, they also bump into each other and they exert forces other than those of impact. Let's simplify the model material by imagining that it is composed of molecules that do not interact with each other, do not even really know about each other and can only bump into the boundaries of the space in which they occur—the walls of the container in which they are placed.

Does such a model even make sense? It surely does, it just couldn't apply to liquids or solids, where the idea of non-interacting molecules is really too constraining. But it could model just gases where the molecules are quite far apart and might not interact. So let's call this model the *ideal gas model*. In fact, this is roughly the idea that Daniel Bernoulli used in the 18th century to explain just the pressure of gases as a result of collisions between particles between which no forces are acting [7]. And why do we want to model anything at all, to look for a mathematical expression? Well, because of the generalization of experimental experience, the possibility to predict even behaviour not observed experimentally (for example, because of the time required for experiments) or even unavailable, because of the ability to design devices and machines using the modelled behaviour. In thermodynamics, Sadi Carnot[17] put it in the words of: to know in advance all the effects of heat acting in a given way on a body [8].

[17] Nicolas Leonard Sadi Carnot (1796–1832), a French military engineer, son of Lazare Carnot, was interested in the theory of the steam engine; his related work, Considerations on the Motive Power of Fire, published in 1824, is considered one of the cornerstones of thermodynamics.

4.3.2 *Mathematical Description of Particle Motion and Impacts*

Let's describe Bernoulli's model of the pressure of an (ideal) gas. We have already mentioned some pressure issues. Thus Bernoulli imagined gas pressure as the result of gas molecules impacting on the walls of the container in which the gas is enclosed. The impacts are the result of the (just thermal) motion of the molecules, and its characteristic called momentum determines, according to Newton's law, the force acting at impact. Bernoulli considered a gas in a vessel in a "final", steady state, otherwise of no movement. For example, a steel cylinder with gas (where the pressure is indicated by a connected pressure gauge) or an inflated balloon, which the pressure of the gas together with the strength of the walls formed into a nicely spherical ball. Pressure can be understood and defined as that impact force per unit area on which it acts. Thus, when we know the applied force(s) and the corresponding area, we know pressure.

Finally, let's start modelling mathematically [9]. The relationship between the momentum of the impacting molecule and the average (constant) impact force is given by Newton's law:

$$\vec{F} = \mathrm{d}(m\vec{v})/\mathrm{d}t \equiv \mathrm{d}\vec{p}/\mathrm{d}t \Rightarrow \vec{F}\Delta t = \Delta\vec{p} \tag{4.9}$$

If we observe particle impacts for a certain time Δt, the total impact force will be proportional to the momentum imparted by the impact $\Delta\vec{p}$. Let us try to express this change in particle momentum on impact in more detail. Bernoulli assumed that the particles are completely elastic. This means that after the impact they move with the same velocity as before, but in the opposite direction. In the words of symbols:

$$\Delta\vec{p} \equiv \vec{p}_{\text{after}} - \vec{p}_{\text{before}} = -m\vec{v} - m\vec{v} = -2m\vec{v} \tag{4.10}$$

For the sake of simplicity, let's first examine only a single particle, and let's concentrate on the vector quantities in the direction of, for example, the x-axis. Thus, we will also be interested only in the magnitude of the vector, or this component of it; we will denote it by a simple symbol, without an arrow, e.g.: $|\vec{p}| \equiv p$. Thus, according to Eq. (4.10) we have for one particle (of mass m): $\Delta p_{x,1} = 2mv_x$. However, the pressure of the gas creates the impact of all the particles. So how many particles actually collide in the x-axis direction in the time we are concerned with (Δt)? Well, all the ones that aren't too far away, that make it to the wall. And if they have a velocity of v_x, they will travel a trajectory of at most $v_x\Delta t$. So all particles that are this far away from the wall or closer will hit, see Fig. 4.2. If the impact area has a size A, this area and the maximum possible particle distance will define a volume of size $v_x\Delta t A$ (see Fig. 4.2). Thus, all particles occurring in this volume hit the surface A and contribute to the total pressure value. Really all of them? After all, there are certainly particles in that volume that fly away from the wall, perhaps after their impact before the time Δt. If all the particles move completely randomly, we can

Fig. 4.2 A model of the particle motion of an ideal gas

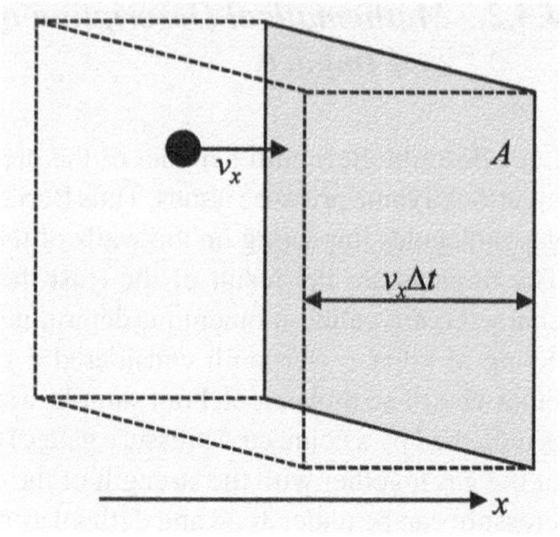

assume that on average half of our volume travels towards the wall and half travels away from it. Otherwise, some direction of motion would be preferred and the gas would flow in that preferred direction, a case we are not interested in now. We're dealing with a pretty round balloon now.

4.3.3 The Pressure of the Ideal Gas

So now we know where the particles are that we have to half-count; but how many are there? We can only find out in general terms—let's say that there are N_V particles per unit volume of gas. Then we get the total number of impacting particles (N_x) by simple multiplication:

$$N_x = N_V(v_x \Delta t A)/2. \tag{4.11}$$

The change in momentum due to all particles is again obtained by simple multiplication:

$$\Delta p_x = N_x \Delta p_{x,1} = N_V(v_x \Delta t A) m v_x. \tag{4.12}$$

Finally, the magnitude of the pressure, which is not directionally dependent, being the same in all directions, is:

$$P = F_x/A = \Delta p_x/\Delta t A = N_V(v_x)^2 m \tag{4.13}$$

Assuming again a completely random motion, i.e., equally likely in the direction of all three coordinate axes, the magnitudes of the three velocity components will be equal: $v_x = v_y = v_z$ and for the magnitude of the velocity vector itself we get

$$v \equiv |\vec{v}| = \sqrt{(v_x)^2 + (v_y)^2 + (v_z)^2}$$

and when substituted into Eq. (4.13)

$$P = (1/3)N_V v^2 m = (2/3)N_V e_k \tag{4.14}$$

where e_k is the kinetic energy of one gas particle, i.e. $(1/2)mv^2$. We thus conclude that the pressure of the gas in our model is given by the kinetic energy of the particles that make it up. Using this simple model, we have thus derived a relationship between gas pressure and the characteristics (velocity) of the microscopic motion that causes it. We have thus revealed the connection between microscopic and macroscopic behaviour or insight.

Note, however, that during the derivation we tacitly introduced another simplification. We assumed that the velocity of all particles is the same, which is certainly not true in reality. In general, the velocity of each particle may be different and their values will be in some interval of real numbers. In fact, we have been working with some average, mean value of the velocity (and associated force, as we have already subtly hinted above Eq. (4.9)). We further assumed that the square of this mean value was equal to the average (mean) of the squares of the individual velocities.

4.3.4 Kinetic Energy of Particles and Temperature

What is actually the relationship between Eq. (4.14) and that which follows for pressure from Eq. (4.8)? The former was derived purely from theoretical considerations, the latter from experimental experience. Equation (4.8) is called the *ideal gas equation of state*. Why an ideal gas? Although the equation was derived on the basis of experimental data, i.e., on the basis of the behaviour of real gases, it was soon found that its validity in the real world was limited—to low pressures and reasonably high temperatures, i.e., to some ideal conditions. It varies from gas to gas how ideal the behaviour is, examples are shown in Fig. 4.3. The ideal gas is thus the simplest model of a gas: there are no interactions between its particles, its particles are "unaware" of each other, they do not interfere with each other's motion, and the entire volume of the gas container is available to it. These are in fact the assumptions of our (i.e., Bernoulli's) simple micromodel that led to Eq. (4.14). Let us therefore combine this equation with the expression $P = nRT/V$ resulting from Eq. (4.8), using Avogadro's constant (N_A):

$$P = N_V RT/N_A = (2/3)N_V e_k \Rightarrow e_k = (3/2)kT \tag{4.15}$$

where $k = R/N_A$ is the Boltzmann constant.

We came up with an interesting finding—the temperature (at least in the case of an ideal gas) is directly determined by the kinetic energy of the particles. Temperature can thus be understood as a measure of the kinetic energy of the particles forming the body (at least for an ideal gas body), the energy of the motion mentioned in Sect. 3.6. Experimental experience teaches us that the pressure of a gas in a given

Fig. 4.3 Behaviour of real gases

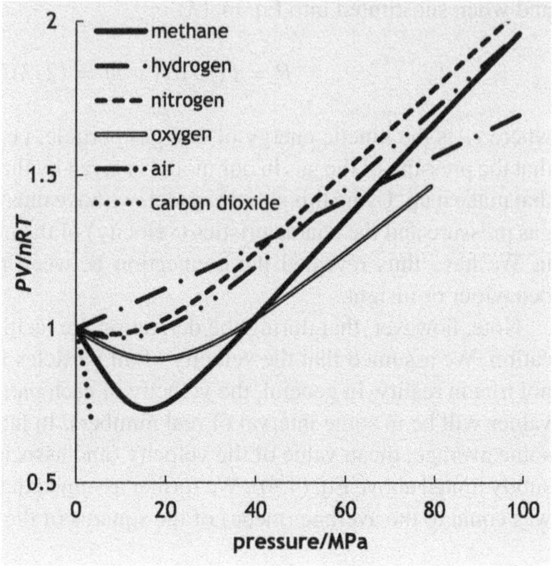

container (in a given volume) depends on its temperature (in addition to how much gas we put in the container). The higher the temperature, the higher the pressure—at least in the experiments we described above. Our simple "micromodel" again shows the relation of pressure to the speed of motion of the gas particles—the higher the speed, the higher the pressure. And it also shows the relationship of temperature to this velocity and the associated kinetic energy of the gas molecules; the higher the temperature, the higher the velocity and kinetic energy.

4.4 Heat and (Transfer of) Energy

This definitively answers the question about the meaning of the micromodel posed a few paragraphs above (Sect. 4.3.1), and, as a bonus, gained another idea of what temperature is. And we can also finally return to the "communication" between hot and cold bodies and generalize the knowledge about the behaviour of gases. The particles that make up a hot body will have higher (kinetic) energy, they will be more mobile than the particles of a cold body. On impact, the more mobile particles will transfer some of their motion, their energy, to the particles of the cooler body. This will raise their temperature or perhaps release them from the crystal lattice and cause them to melt. So when a hot and a cold body come into contact, their molecules transfer, exchange energy through their collisions. We call this exchange, the transfer of energy at the microscopic level, heat at the macro level and assign it a value to determine the energy exchanged.

Thus, *heat* is the name for the way—process—by which bodies of different temperature can exchange, transfer energy. So heat is not something that is "contained" in a body and can be "transferred" to another body—energy is still being "transferred". Heat refers to the process, the method of energy transfer, which we discussed in Sect. 3.5. The number assigned to heat indicates the amount of energy transferred and therefore has a unit of energy.

The concept of heat has a similar meaning to the concept of walking. The latter denotes a way of moving from place to place and is characterized by the trajectory that has been travelled in doing so. Walking is not something that is in the feet or getting into them, it is the name of a type of motion. As a motion, walking connects the starting and ending points, which can be characterized by coordinates in a network of meridians and parallels, as for example GPS can do. How many kilometres we travel between the starting and ending points depends not only on the positional coordinates and the straight-line distance between the two places, but also on which of the many possible paths and routes we take.

In the case of heat, the starting and ending points may be the starting and ending temperatures of the body or bodies that are of interest to us. The total energy exchanged by heat depends on the "heating and cooling path" between these points. However, the difference in energy between the starting and ending points is determined only by these points, and does not depend on that path.

Heat is closely related to microscopic—molecular, atomic—motion and to its energy as one of the fundamental characteristics of motion. Let us recall one important, fundamental characteristic of energy—it is a quantity that is conserved, it does not disappear or appear out of nowhere. But it can move from body to body (from system to system). If this is happening and we focus only on the body itself, it can seemingly lose or gain energy. This is where energy exchange processes such as heat come in. If the exchange (transfer) of energy is by heat, we have to take the numerical value of heat into account when examining the conservation of energy. If, for example, a body loses energy by heat, the sum of the value of its current energy and the value of the heat gives its initial energy; nothing is lost. However, the body cannot be assigned a heat value.

Let us not confuse heat with forms of energy such as kinetic and potential energy. These too must be included in the conservation of (total) energy, especially if they can interpenetrate; the individual forms of energy need not themselves be conserved. They are not, however, process quantities like heat.

Conservation of energy is not a trivial matter and we must be careful not to make a mistake. We have to notice the different forms of energy and the boundaries of the body or system whose energy we are interested in—whether they are "impermeable" to energy exchange, or whether it can occur, for example, through heat.

4.5 First Experience with Heat Measurement

4.5.1 Black and Contemporaries

Let's still stay with heat and talk about the (history of) its measurement. The measurement of physical quantities is often an illustrative aid to understanding their nature, meaning and reason for introduction. We use the concept of heat quite commonly in everyday life. People have perhaps always intuitively associated a drop in temperature with a loss of something they called heat, and an increase in temperature with a gain of heat. For a long time, too, the caloric theory of heat lived, and almost reigned, regarding heat as a really material, though weightless substance; but we will not dwell on it here, because today, let us repeat, we understand heat as a process, not as a substance, as one of the modes of transfer of energy between bodies.

The more exact measurement of heat (we would prefer to say the measurement of the energy transferred by heat, but we follow this abbreviation in this chapter, as it has become established in the literature) is precisely connected with the invention of the *thermometer*. This was the only instrument that allowed to measure, to quantify our thermal experience, not only in terms of the state of a body—warm, cold, lukewarm, freezing—but also, to some extent, in terms of the change in these states, which are precisely the result of the transfer of energy by heat. Probably the first measurements in this direction come from simple mixing experiments [10]. Boerhaave describes those perhaps carried out by Fahrenheit.[18] He mixed equal quantities of hot and cold samples of the same liquid (he tried water, vinegar, alcohol, oil) and found that the resulting temperature always lay midway between the original temperature of the hot and cold liquid. But when, at Boerhaave's request, he made the same mixing with water and mercury, he found that this rule did not apply. The resulting temperature was closer to the initial water temperature. Joseph Black[19] explained the result by the smaller "heat capacity" of mercury than water. Black was one of those who contributed substantially to the understanding of the differences between temperature and heat and the nature of the phenomena in which temperature changes with heat.

Black used mixing experiments to estimate these capacities—he mixed two bodies of equal mass but different temperatures, their "capacities" were then inversely proportional to the change in temperature of each. So, for example, he mixed a pound of gold at 150° with a pound of water at 50° and found that the resulting temperature (of the water with the gold) was 55°. Gold seems to have lost 95°, while water seems to have gained 5°. The capacity of gold to the capacity of water of the same weight is therefore in the ratio 5–95, i.e., 1–19. The mixing experiments were thus based on the change in temperature and assumed that this change was due to "heat exchange"

[18] Daniel Gabriel Fahrenheit (1686–1734), a Dutch physicist and inventor of German origin, born in Poland. He constructed meteorological instruments, invented the mercury and alcohol thermometer. Their original scale is still used in some countries under his name.

[19] (1728–1799), a Scottish physician and chemist born in France, professor at the universities of Glasgow and (later) Edinburgh, rediscoverer of carbon dioxide.

(we would say due to the transfer of energy by heat). The (energy) loss by heat is manifested by a decrease in temperature, the heat gain by an increase in temperature. Black, on the basis of his experiments, argued that equal masses of different bodies require different heats for the same temperature rise—he called them capacities for heat, in today's terms, heat capacities.

4.5.2 The First Quantifications of Heat

The first table of heats, called specific heats, was published by Magellan[20] in 1780 [10]. However, at that time, specific heat did not mean the same thing as it does today, i.e., specific heat capacity, but the (assumed) amount of heat contained in a unit mass of a body in a specified state. If the specific heat of water is taken there as equal to one, then, for example, ice has a value of 0.9, mercury 0.033, iron 0.125, tin 0.068, lead 0.050, air 18.67, carbon dioxide 0.270. Here we are not evaluating the reality of these figures from the level of present knowledge, but are giving them as an illustration of the fact that they are relative values, relative to water as the chosen unit.

But Black noticed other things that he associated with warmth, too. He wondered why it takes a long time for all the winter snow and ice to melt after winter. It doesn't disappear at once when the temperature rises above 0 °C. Black was convinced that it takes quite a lot of heat to melt snow or ice (the amount of energy supplied by heat, we say here) and it takes quite a long time to be "extracted" from the surrounding environment of snow, ice. He had similar thoughts about the evaporation of water (liquids) and in both cases he supported them with experiments. Today, thanks to Black, we talk about heat of melting or of evaporation.[21]

The basis of Black's experiments was therefore the measurement of temperature changes—how ingeniously he did this in the case of melting or evaporation when boiling, when the temperature does not actually change, the inquisitive reader will learn, for example, in the book [10]. Here it is sufficient to emphasize that Black detected heat according to how many degrees (Fahrenheit in his case) the temperature of water of the same mass as the body whose thermal behaviour, whose energy exchange by heat, he was investigating. Heat was thus measurable by its effect on temperature. It is actually a double mediated measurement—already temperature itself is measured by some other property of the body, and heat is then measured by temperature (its changes). In fact, that has remained to this day. In Black's and slightly later times, water was a kind of standard for measuring heat in this way.

[20] Jean Hyacinthe de Magellan, in Portuguese João Jacinto de Magalhães (1723–1790), a Portuguese naturalist, a direct descendant of the famous seafarer, exchanged the monastic life for a scientific one at the age of 40 and settled in London.

[21] Black himself talked about latent heat.

Lavoisier[22] and Laplace[23] used a different principle to measure heat [11], though still based on water (they conducted their first experiments according to Fenby [12] in the winter of 1782 and published in 1783–1784). Their ice calorimeter contained ice, which was (partially) melted by the exchange of energy through heat, and the resulting liquid water was captured and weighed. The mass of the resulting liquid water was proportional to the heat detected. Knowing the heat of melting of the ice, it could be converted directly into units of heat. The ice calorimeter also allowed to measure thermal effects of chemical reactions. Lavoisier and Laplace, for example, wrote that the explosion (reaction) of 1 ounce of potassium nitrate with 1/3 ounce of carbon would melt 12 ounces of ice [13].

4.5.3 More Experiments with Heat—Joule

Other experiments, important for our exposition, were carried out by Joule,[24] who was not specifically concerned with the methodology of heat measurement, but with something a step further. He was trying (among other things) to measure the heat associated with macroscopic, mechanical motion, which we know from common experience as frictional heating. Energy, exchanged between bodies by the force of mechanical motion, is transferred and measured by work, another important quantity for us, to which we shall return in due course. A reminder from elementary physics will suffice for now—work is determined by the force (as the cause of motion already known to us) and the path along which it acts, along which the corresponding motion takes place, by the product of the force and the path length. It is customary to name the exchange of energy by work, somewhat inaccurately, doing work. To put it another way, Joule was interested in the heat associated with the doing work. Indeed, the temperature of a body can be increased not only by heating (heat; due to contact with another body of higher temperature), but also by mechanical motion, or, more simply, by work [14]. If the temperature change is the same in both cases, we can say that heat and work are equivalent here. In other words [15], if work is accompanied by friction, there is a change in the temperature of the body in question, even if it is not in contact with a body of a different temperature. But in doing so, we can determine or calculate the heat required to produce the same temperature change—in this sense,

[22] Antoine Laurent Lavoisier (1743–1794), a French chemist and also a lawyer or tax official, founder of modern chemistry, member of the Academy of Sciences, from 1780 Fermier Général, something like a royal tax collector, executed during the Great French Revolution.

[23] Pierre Simon de Laplace (1749–1827), a French mathematician, physicist and astronomer, also involved in politics, one of the greatest natural scientists ever, probably best known for his studies of the stability of the solar system.

[24] James Prescott Joule (1818–1889), an English brewer, yet mainly a physicist- and experimenter-privateer, a pupil of John Dalton, supported by Kelvin, a member of the Royal Society from 1850.

mechanical and thermal actions are equivalent (i.e., work and heat are equivalent). For they have a common denominator—energy—they are both possible processes of energy transfer between bodies.

In 1845, Joule published the results of his experiments to determine the temperature changes during condensation (which meant compression) and dilution (which in turn meant expansion) of gases [13]. The essence of his experiments was a gas reservoir coupled to a pump—designed simply as a cylinder with a piston enclosing the gas—both immersed in a water bath whose temperature was measured by a sensitive thermometer. From the dimensions and motion of the pump, i.e., the piston, and the force exerted, he ascertained the work involved in heating the water. He found that to raise the temperature of the water by one degree Fahrenheit required a work equivalent to lifting 823 pounds to a height of one foot.

Further experiments were carried out by Joule with a paddle wheel (roughly between 1845 and 1850) immersed again in a water bath, which was connected to a weight via a rope gear. The dropping of the weight due to gravity caused the wheel to spin, the blades cut through the water and Joule noted the rise in its temperature. From the mass of the weight and the trajectory it travelled, he determined the work associated with its motion. From a series of experiments, also carried out with mercury or sperm oil, he concluded that the heat associated with friction for all bodies is proportional to the work "done". To raise the temperature of a pound of water (having a temperature between 55° and 60°) by 1 °F requires a "mechanical force" (work) representing a drop of 1 foot in a weight of 772 pounds.

4.5.4 Again to the Heat Quantification; Calories

Joule's standard of measurement for heat was also water. We can take the heat causing the temperature of water (a chosen quantity) to change by one degree to be unit of heat and call it a *calorie*. From Joule's data we can see that one calorie is equivalent to 823 pound-feet for gas compression experiments and 772 pound-feet for paddle wheel experiments. These are close values, but again far enough away to see a universal equivalence of energy transfer by heat and work. Unfortunately, Joule's experiments were not characterized by high accuracy—his measured equivalent values ranged from 742 to 1040 pound-feet, and he reportedly mentions one experiment with a result of 587 pound-feet. This was also the reason for the reluctance of some of his contemporaries to accept them [8]. We will not discuss this difficulty further here, and we will allow ourselves at this point to make a small leap by the simplified statement that it is also thanks to Joule that we take the relation between work and heat, or their units, to be established: $[W] = J[Q]$ (the square brackets symbolise the units of the quantity enclosed in them). The constant J just interconverts the units of work and heat. For Joule, it would be equal to, say, 772 lbf/cal. Today we express work in joules, and until recently, we used the calorie as the unit of heat.

In this case $J = 4.185$ J/cal. However, nothing prevents us from choosing $J = 1$ in that relationship and measuring heat and work in the same units. That is what we are doing today, in joules.

Here we must still make a comment on the calorie as a unit of heat. Its foundation on water heating is complicated by the fact that the heat involved depends on the (initial) temperature of the water. Originally Regnault defined a calorie for heating 1 g of water from 0 to 1 °C, later it was a hundredth of a heat for heating (again 1 g) from 0 to 100 °C and finally and most firmly, the definition of the so-called 15° *gram-calorie* (g.cal.) was based on heating 1 g of water from 14.5 to 15.5 °C (abbreviated and simply denoted cal).

References

1. Rajagopal, K.R.: Remarks on the notion of "pressure". Int. J. Non-Linear Mech. **71**, 165–172 (2015)
2. Cohen, I.B.: Newton, Hooke, and "Boyle's Law". Nature **204**(4959), 618–621 (1964)
3. Richmond, T.G., Parr, A.: A U tube experiment to discover the curve in Boyle's law. J. Chem. Educ. **74**(4), 414–415 (1997)
4. Partington, J.R.: A History of Chemistry, vols. 2, 3. MacMillan & Co., London (1961, 1962)
5. Monk, P.: Physical Chemistry: Understanding our Chemical World. Wiley, Chichester (2004)
6. Pekař, M., Samohýl, I.: The Thermodynamics of Linear Fluids and Fluid Mixtures. Springer, Cham (2014). See also https://www.fch.vut.cz/vav/sapelite
7. Partington, J.R.: An Advanced Treatise on Physical Chemistry. The Fundamental Principles - The Properties of Gases, vol. 1. Longmans, London (1949)
8. Truesdell, C.: The Tragicomical History of Thermodynamics 1822–1854. Springer, New York (1980)
9. Pekař, M.: Lecture Notes on Thermodynamics and Electrochemistry. Brno: Brno University of Technology, Faculty of Chemistry (2008)
10. McKie, D., Heathcote, N.H. de V.: The Discovery of Specific and Latent Heats. E. Arnold & Co., London (1935)
11. Armstrong, G.T.: The Calorimeter and its influence on the development of chemistry. J. Chem. Educ. **41**(6), 297–307 (1964)
12. Fenby, D.V.: Heat: its measurement from Galileo to Lavoisier. Pure Appl. Chem. **59**(1), 91–100 (1987)
13. Lewis, C.J.T.: Heat and Thermodynamics: A Historical Perspective. Greenwood Press, Westport (2007)
14. Greenslade, T.B.: Nineteenth-century measurements of the mechanical equivalent of heat. Phys. Teach. **40**(4), 243–248 (2002)
15. Arons, A.B.: Development of energy concepts in introductory physics courses. Am. J. Phys. **67**(12), 1063–1067 (1999)

Chapter 5
Back to the Thermal Behaviour of Gases—Real Gases

Abstract Deviations from Boyle's law, particularly in Andrews' measurements are presented. The pathway to the Van der Waals equation of state is outlined. The critical state is mentioned and reduced quantities are introduced.

Until the late eighteenth century, Boyle's law was (successfully) applied only to air. In 1799, van Marum,[1] probably provided the first reliable data on deviations from Boyle's law, in the case of ammonia under high pressure, which led to its liquefaction. He showed that here the product of pressure and volume decreases with increasing pressure, i.e., that the gas is more compressible (has less volume at a given pressure) than Boyle's law would imply. Similar behaviour was subsequently found for other gases [1], whether easily (hydrogen sulphide, cyanogen) or only very difficult (nitrogen, hydrogen, ethylene, etc.) to liquefy, in the works of Regnault, Natterer,[2] Cailletet,[3] and especially Amagat.[4] However, Natterer was the first to find, in addition, that the product of pressure and volume after an initial decrease with increasing pressure begins to increase again (for all the gases studied). They all used a suitable liquid, usually mercury or water, as the "pressure source", so it was still essentially hydrostatic pressure.

Historically, perhaps the most famous first thorough experiments with the behaviour of gases at higher pressures, leading to their liquefaction, were carried

---

[1] Martinus van Marum (1750–1837), a Dutch physician, natural scientist, designer of demonstration instruments, founder of the Teyler Museum in Haarlem.

[2] Johann Natterer (1787–1843), an Austrian naturalist and traveller, participant in the Imperial Expedition to South America (Brazil) in 1817–1835.

[3] Louis-Paul Cailletet (1832–1913), French physicist, inventor, metalworker, known for his work on the liquefaction of gases.

[4] Émile Hilaire Amagat (1841–1915), a French physicist, studied the isotherms of many gases even at very high pressures, for which he invented the hydraulic manometer. The law on partial volumes of gases bears his name.

© The Author(s), under exclusive license to Springer Nature Switzerland AG 2024

M. Pekař, *The Essentials of Thermodynamics*, Physical Chemistry in Action,
https://doi.org/10.1007/978-3-031-60321-1_5

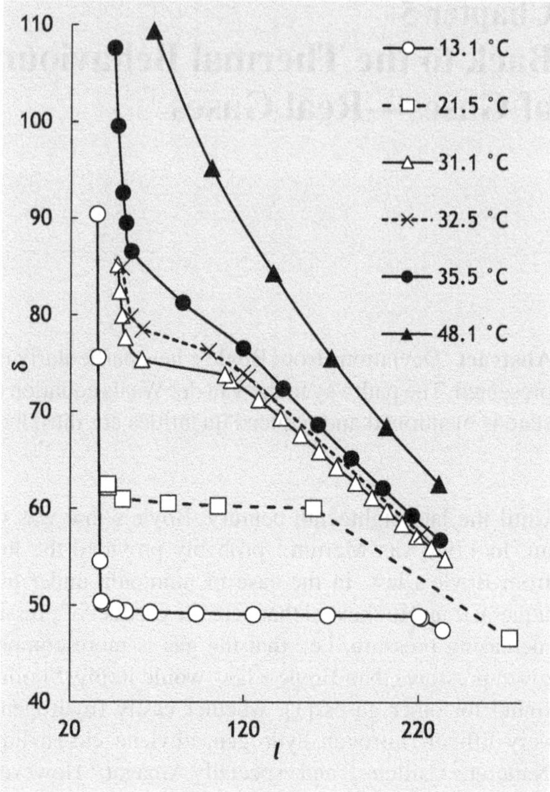

Fig. 5.1 Data from Andrews measurements—carbon dioxide behaviour

out by Andrews[5] [2]. He started with carbon dioxide,[6] which he effectively compared with air, which served as an indicator of pressure. His experimental apparatus consisted in principle of a glass capillary sealed at one end, in which the gas was enclosed by a column of mercury. The capillary was fixed in a sealed copper cylinder filled with water, which could be compressed by means of a steel piston fixed in a tight fitting; this varied the pressure in the capillary. Andrews used two "synchronous" cylinders, one containing a capillary containing CO_2, the other a capillary containing air. The cylinders were tempered to stable temperatures; by measuring the volume of air in the capillary, he could determine the applied pressure using Boyle's law (after temperature conversion of Charles' law). His pressure was therefore, again in principle, hydrostatic pressure.

Andrews' original data can be seen in Fig. 5.1 in his original coordinates, where δ represents pressure and l represents volume (Andrews used relative quantities, see his original paper for details). More recent data over a relatively narrow range of

[5] Thomas Andrews (1813–1885), an Irish chemist and physicist, professor at Queen's College in Belfast.

[6] These measurements were later extended just by Amagat.

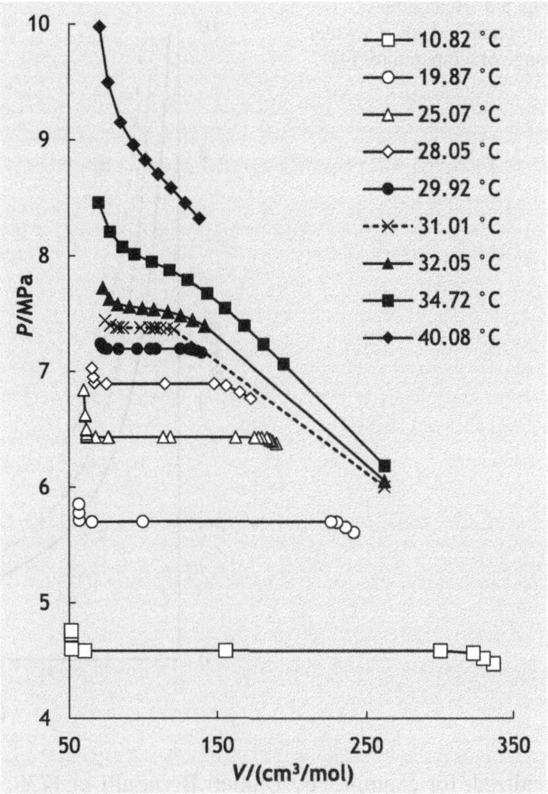

Fig. 5.2 More recent data on carbon dioxide behaviour [3]

temperatures is shown in Fig. 5.2. All the data clearly show not only the behaviour of carbon dioxide gas at lower temperatures, but also the transition to liquid form and its behaviour. The data for just carbon dioxide gas over a wider temperature range are shown in Fig. 5.3. Figure 5.4 then shows a comparison with the behaviour of an ideal gas for a single temperature.

5.1 More Realistic Equations of State; Van der Waals

Experimenting with gases in retrospect has led to somewhat contradictory consequences. It was the basis for the formulation of the simple Boyle's law and later the Charles (Gay-Lussac) law, which we combined in Eq. (4.8). At the same time, however, (more thorough and especially over a wide range of pressures) experiments soon pointed out deviations of the behaviour of real gases from these laws, or from their combined form. Further, the experiments also provided guidance on where to look for one of the two fundamental causes of these deviations. The first was already

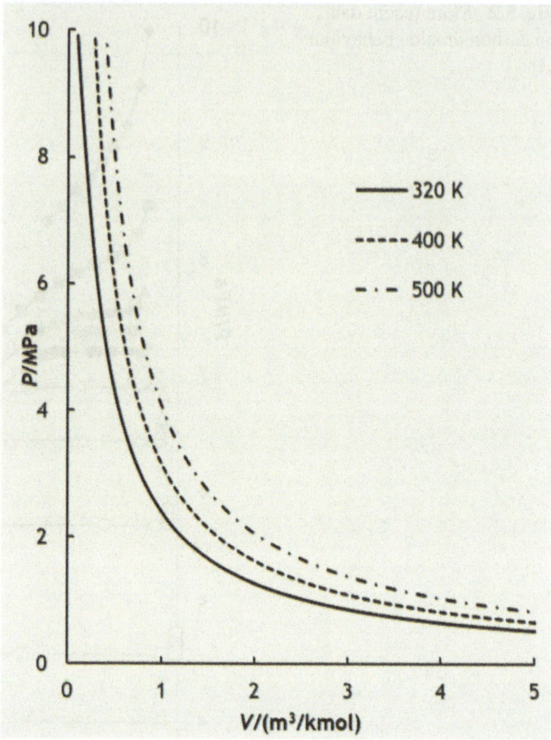

Fig. 5.3 Behaviour of carbon dioxide over a wider range of temperatures [4]

realized, for example, by Daniell Bernoulli in 1738, when he derived Boyle's law theoretically—gas molecules themselves have a certain volume, and the volume (of the container) appearing in the equations of that law must be corrected for that volume. In other words, the gas molecules (particles) do not have the whole volume V, available to them, but only a part of it, the "free volume", which they do not themselves occupy.

In experiments on the liquefaction of gases by increased pressure, it has been found that the volume of the liquid changes very little even with a considerable increase in pressure. This, on the whole, easily accepted experience points to strong repulsive forces between molecules in the liquid state, preventing their further approach. In the liquid state, the molecules are certainly much more closely arranged than in the gaseous state. However, the phenomenon of liquefaction of gases by increased pressure also indicates the existence of attractive interactions between molecules, which become stronger as the volume of the gas decreases. In short, the second reason for the deviation of the behaviour of real gases from the model (4.8) is the existence of (intermolecular) interactions.

Thus, the idealized mathematical description (model) in the form of Eq. (4.8) too often ceases to be valid for real gases. During the second half of the nineteenth century, attempts were made to design (state) equations taking into account deviations from the ideal gas equation of state, but none of them caught on significantly, probably

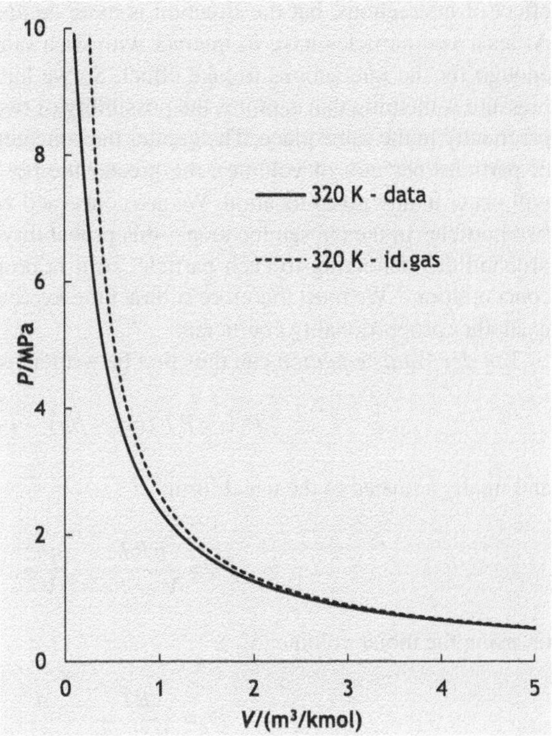

Fig. 5.4 Comparison of the behaviour of carbon dioxide [4] and ideal gas

for their overly empirical approach. It was not until the equation proposed in 1873 by Van der Waals[7] that it literally became famous and survives to this day as the first and simplest equation of state for a real gas. Famous probably because it also made it possible to describe the liquid state—in short and simplified terms, the "whole" curves measured by Andrews. Let us explain in a rough outline the considerations that led to the derivation of the Van der Waals equation.

The first step is the aforementioned correction of the volume to the actual volume of the particles (molecules or atoms) that make up the gas. For different gases we must subtract different numbers from the volume V of the container, but in general it will always be proportional to the molar amount of gas (number of its particles) that we have enclosed in the volume V. If we denote the gas molar amount n, the coefficient of proportionality by b, then the corrected volume is equal to $V - bn$.

Next, we must correct the pressure for intermolecular interactions. These "hinder" the motion of the particles and result in a reduction of their impact on the wall, i.e., a reduction in pressure (compared to the ideal). The more particles, the greater the

[7] Johannes Diderik Van der Waals (1837–1923), a Dutch physicist, professor at the University of Amsterdam. His work on the behaviour of gases and liquids was awarded the Nobel Prize for Physics in 1910. Some of the intermolecular interactions co-responsible for non-ideal behaviour still bear his name.

effect of interactions, but the situation is more complex for these than for volume. At least two particles have to interact with each other and they have to get close enough for the interactions to take effect. So we have to subtract from the (ideal) pressure something that captures the possibility of two particles very close together, practically in the same place. The greater the concentration of particles (the amount of particles per unit of volume), the greater the possibility (probability) that they will occur at any given location. We are concerned with the occurrence of actually two particles in the chosen location—this probability is given by the product of the probabilities belonging to each particle, so it is proportional to the square of the concentration.[8] We must therefore subtract the expression $ac^2 = a(n/V)^2$, where a is another proportionality coefficient.

Van der Waals equation can thus first be written as follows:

$$P = nRT/(V - bn) - ac^2$$

and finally adjusted to the usual form:

$$P = \frac{nRT}{V - bn} - \frac{an^2}{V^2}, \quad (5.1)$$

or, using the molar volume (V_m):

$$P = \frac{RT}{V_m - b} - \frac{a}{V_m^2}. \quad (5.2)$$

We pay a price for approximating reality in that the constants a and b are not universal but must be determined for each gas. Some examples of their values are given in Table 5.1. But the Van der Waals equation has one universal mathematical feature, which, moreover, corresponds to a universal phenomenon already observed, for example, by Andrews. The isotherms calculated from the Van der Waals equation have local minima and maxima in the pressure-(molar) volume coordinates at low temperatures. As temperature increases, these extremes approach each other, merge at a certain temperature, and no longer appear above that temperature. An example for carbon dioxide is shown in Fig. 5.5. The convergence creates an inflection point in the coordinates shown on the Van der Waals curve. The temperature at which this occurs is called the critical temperature and corresponds to the highest temperature at which the gas can still be liquefied by increasing the pressure. The inflection point represents a *critical state* which, in addition to a certain temperature, has a certain, specific value of critical pressure and (molar) volume. If Eq. (5.2) is understood as

[8] Example for better understanding: if a dice is rolled, the probability of rolling any of the six numbers is equal to 1/6 for each number; if two dice are rolled simultaneously, the probability of rolling some pair of numbers is equal to (1/6)(1/6) = 1/36; 36 is also the number of all possible pairs produced by the combination of the two dice.

Table 5.1 Examples of Van der Waals equation parameter values [5]

Gas	a 10^5Pa(m^3/kmol)^2	b m^3/kmol
Nitrogen	1.361	0.0385
Oxygen	1.369	0.0315
Methane	2.285	0.0427
Carbon dioxide	3.643	0.0427
Hydrogen	0.247	0.0265

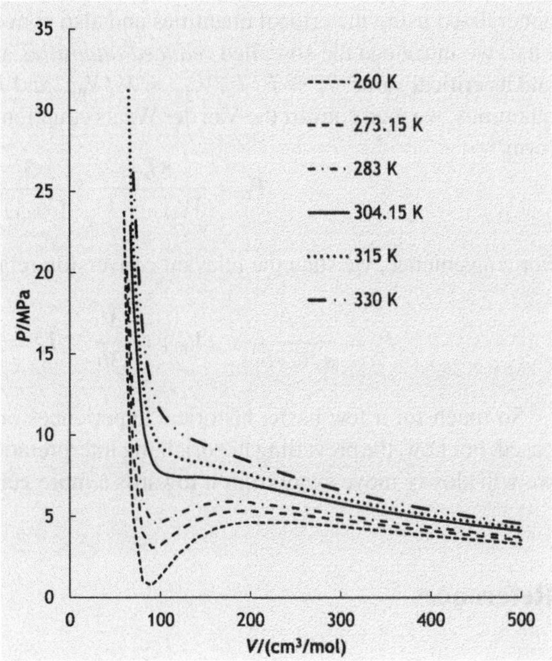

Fig. 5.5 Van der Waals equation for carbon dioxide

an expression of the function $P = \hat{P}(T, V_m)$, then at the inflection (critical) point (subscript c) of this function the following must be valid:

$$\frac{\partial \hat{P}}{\partial V} = 0 = -\frac{RT_c}{(V_{m,c} - b)^2} + \frac{2a}{(V_{m,c})^3}, \tag{5.3}$$

$$\frac{\partial^2 \hat{P}}{\partial V^2} = 0 = \frac{2RT_c}{(V_{m,c} - b)^3} - \frac{6a}{(V_{m,c})^4}. \tag{5.4}$$

Combining these expressions with Eq. (5.2) we can finally express the constants a and b using critical quantities, e.g.,

$$a = \frac{9}{8} R T_c V_{m,c} = \frac{27 R^2 T_c^2}{64 P_c}, \tag{5.5}$$

$$b = \frac{1}{3} V_{m,c} = \frac{R T_c}{8 P_c}. \tag{5.6}$$

For example, carbon dioxide has a critical temperature of 31.1 °C and a critical pressure of 7390 kPa.

The critical state occurs generally for any substance (gas), albeit at different values of temperature, pressure and (molar) volume. The Van der Waals equation can thus be generalized using the critical quantities and also converted to a dimensionless form. First, we introduce the so-called *reduced quantities* as the ratio of a given quantity and its critical value: $T_r = T/T_c$, $V_{m,r} = V/V_{m,c}$ and $P_r = P/P_c$. Using the reduced quantities, we then convert the Van der Waals equation into a reduced, dimensionless form

$$P_r = \frac{8T_r}{3V_{m,r} - 1} - \frac{3}{(V_{m,r})^2}. \tag{5.7}$$

For convenience, we state the relevant conversion relations explicitly:

$$P_r = \frac{P}{a/(27b^2)}, \quad V_{m,r} = \frac{V}{3b}, \quad T_r = \frac{T}{(8a)/(27Rb)}.$$

So much for a few basic, historical experiences on which thermodynamics was based. For now, the prevailing historicizing interpretation will suffice for the moment; we will slowly move away from it towards a more generalizing, theoretical insight.

References

1. Partington, J.R.: A history of chemistry, vols. 2, 3. MacMillan & Co., London (1961, 1962)
2. Andrews, T.: Über die Continuität der gasförmigen und flüssigen Zustände der Materie. Leipzig: W. Engelmann, 1902 (German translation of English original)
3. Michels, A., Blaisse, B., Michels, C.: The isotherms of CO_2 in the neighbourhood of the critical point and round the coexistence line. Proc. R. Soc. Lond. Ser. A **160**(902), 358–375 (1937)
4. Linstrom, P.J., Mallard, W.G. (eds.): NIST Chemistry WebBook, NIST Standard Reference Database Number 69. National Institute of Standards and Technology, Gaithersburg. https://doi.org/10.18434/T4D303 (citováno 6. řervna 2022)
5. Turns, S.R.: Thermodynamics: Concepts and Applications. Cambridge University Press, New York (2006)

Chapter 6
Thermodynamics—A Basic Macroscopic Model

Abstract The concept of heating as an energy flow is introduced as well as a basic mathematical model of its relationship to changes in the temperature and volume of a body. The roles of temperature and volume as the principal independent variables are stressed. The concepts of the constitutive domain and the constitutive equation are introduced and illustrated. Pressure is formulated as a function of temperature and volume. Processes and their curves are explained mathematically; isobars of an ideal gas, a Van der Waals fluid, and water are examined as examples. The potential role of pressure as an independent variable is mentioned.

6.1 Motivation for Modelling

So far, we have mainly focused on an empirical-historical view of the essentials of thermodynamics. We have seen that the study and description of thermal behaviour, the thermal properties of bodies, is the practical essence of thermodynamics. One of the men of thermodynamic history, Fourier,[1] says [1]: by the action of heat, all bodies—solid, liquid, gaseous—tend to expand[2]; this property makes heat visible. This "visibility" shows us the way from experience to theory, to the fulfilment of the above vision of another founder of thermodynamics, Sadi Carnot: to know in advance all the effects of heat acting in a given way on a body [1]. This means having a suitable (and useful) model at hand that allows us to make such predictions and at the same time can describe that visible experience. Such a mathematical model can thus, on the one hand, handle the knowledge of how much energy must be supplied to a body by heat (transferred from another body) in order to increase its temperature by so many degrees, and at the same time, for example, how its volume will change; on the other hand, it can predict, for example, how the temperature and dimensions

[1] Jean Baptiste Joseph Fourier (1768–1830), a French mathematician and physicist who participated in Napoleon's campaign in Egypt, showed how heat conduction can be described by the infinite series that now bear his name, as well as the related Fourier transform.

[2] Fourier did not have all of today's thermal experience, so he did not know that it is not always about expansion.

© The Author(s), under exclusive license to Springer Nature Switzerland AG 2024
M. Pekař, *The Essentials of Thermodynamics*, Physical Chemistry in Action,
https://doi.org/10.1007/978-3-031-60321-1_6

of a body will change when it is heated or cooled. For this we certainly need to know the characteristics of the given body, for example, how its volume is related to temperature (and possibly other properties), but these must in turn enter into some general (body-independent) model of thermal behaviour.

6.2 A New Concept—Heating as an Energy Flow

Let us summarize the basic experimental findings and experience that a theory, i.e., a general model, should describe. Heating or cooling causes bodies to change at least their temperature or volume[3] (which, especially in the case of gases and liquids, i.e., fluids, also results in a change in pressure). Both heating and cooling involve the exchange (transfer) of energy by heat. This takes place in time, suppose it can be represented (modelled) by a function of time, which we denote by $\hat{Q}_f(t)$. It has the energy·time^{-1} dimension and can thus also be called an energy flux (this is emphasized by the subscript f). If it is an integrable function (over the necessary time interval), and we will assume that it is,[4] then the energy exchanged by heat in a given time interval (Q) can be obtained by integration:

$$Q = \int_{t_1}^{t_2} \hat{Q}_f \mathrm{d}t. \tag{6.1}$$

where $\langle t_1; t_2\rangle$, $t_2 > t_1$ is that interval. To simplify the text, let's call $\hat{Q}_f(t)$ *heating*, whether it is heating or cooling. These two ways are distinguished by the sign of the heating: $Q_f > 0$ means really heating, $Q_f < 0$ cooling. This also reminds us of the current sign convention for energy transfer processes between bodies—the energy that a body gains by transfer has a positive sign, and the energy that a body loses by transfer has a negative sign. The convention is illustrated in Fig. 3.4.

During that time interval $\langle t_1; t_2\rangle$, the body may gain (receive, absorb) energy by heat in some its parts, and lose (give up, emit) energy in other parts; of course, there may also be parts of that interval in which there is no transfer of energy by heat at all. Heating, cooling... all of which can be expressed mathematically by a nonnegative number; namely

$$\begin{aligned} \tfrac{1}{2}(|Q_f| + Q_f) &= \begin{cases} Q_f, & \text{if } Q_f \geq 0, \\ 0, & \text{if } Q_f \leq 0; \end{cases} \\ \tfrac{1}{2}(|Q_f| - Q_f) &= \begin{cases} -Q_f, & \text{if } Q_f \leq 0, \\ 0, & \text{if } Q_f \geq 0 \end{cases} \end{aligned} \tag{6.2}$$

[3] Changes like melting, boiling, etc. are left aside for now.

[4] From the integrability point of view, we need only state the well-known theorem that a piecewise continuous function on a given interval (continuous almost everywhere on it) is integrable.

and then the *energy absorbed* (Q^+) or *emitted by the heat* (Q^-) is obtained as follows:

$$Q^+ \equiv \tfrac{1}{2}\int_{t_1}^{t_2} (|Q_f| + Q_f)\mathrm{d}t \geq 0, \tag{6.3}$$

$$Q^- \equiv \tfrac{1}{2}\int_{t_1}^{t_2} (|Q_f| - Q_f)\mathrm{d}t \geq 0. \tag{6.4}$$

It is clear that Q, Q^+, Q^- have dimensions and units of energy.

6.3 Formulation of a Model; Constitutive Equations

Let us repeat: experience tells us that heating leads to at least a change in volume and temperature. The simplest mathematical model of this experience is probably the following, linear, in the sense of a linear relationship between heating and both volume and temperature change:

$$Q_f = \hat{\Lambda}_V(V, T)\dot{V} + \hat{C}_V(V, T)\dot{T}, \tag{6.5}$$

where parentheses indicate the functional dependence on the variables in parentheses[5] and dots the time derivatives. We shall call the quantity Λ_V the *coefficient of thermal volume change*, and C_V the *heat capacity at constant volume*; as the functions in (6.5) they are continuous and have continuous first derivatives (assumptions of our model). Experience tells us that to raise the temperature of a body without changing its volume, energy must be supplied by heat; we, therefore, add a condition to our model:

$$C_V > 0. \tag{6.6}$$

The unit of heat capacity is apparently $\mathrm{J\,K^{-1}}$, for the unit of the coefficient Λ_V Eq. (6.5) gives $\mathrm{J\,m^{-3}}$, which is the same as Pa—so the coefficient has a pressure dimension.

The model (6.5) is quite general (although its actual validity will certainly be limited); the specific characteristics of particular bodies are introduced by these two coefficients. The specific form of their dependence on temperature and volume may vary from body to body. Finding these dependencies, the equations describing the thermal behaviour for a given body, is one of the tasks of thermodynamics. Thus, these equations are in principle *constitutive* (state) *equations*, i.e., specific to a body,

[5] We will try to differentiate the function symbol, e.g., $\hat{V}$, from the symbol of a function value or quantity, e.g., V; the parentheses immediately after the function symbol thus always denote an enumeration of independent variables. However, we will not do so when the dot—the time derivative—is used and when the function value at a particular point is expressed, e.g., $f(x_1, y_1)$ means the value of the function $\hat{f}(x, y)$ at the point $[x_1, y_1]$.

while Eq. (6.5) is generic, i.e., general for all bodies (for which it is an acceptable model).

6.3.1 Choice of Independent Variables

Our model Eq. (6.5) shows that we have chosen the temperature-volume pair as independent variables. Why? Firstly, for practical reasons—one of our inspirations was our experience with the behaviour of gases, where these two variables are practically the easiest to adjust (by thermostat and vessel of a given volume). Secondly, for historical reasons, which are partly the same and which we will add a little later in the text around the Eq. (6.7). Further, because another possible independent variable—pressure—is not so trouble-free, as we have already discussed and will show. Thus, we will see that the temperature-volume pair is the most appropriate for building a général, if simple, thermodynamic theory as a set of independent variables.

Let us also mention that it is the Van der Waals fluid (equation) that illustrates that pressure (together with temperature) cannot serve as a universal independent variable of a general thermodynamic theory—right here it is not always possible to find a single value of pressure (at a given temperature) in addition to a single value of volume. For volume, this equation is cubic, so it is not always invertible for volume, and for a given pressure (and temperature) it can have up to three different roots; from Fig. 5.5 (or Fig. 6.4 below) it is also clear that below the critical temperature the isotherms intersect the isobars at least at two points. This supports our choice of the volume-temperature pair in role of the fundamental and general pair of independent variables.

6.3.2 Constitutive Domain

However, the volume and temperature generally change with heating itself, and do so over time. We denote derivatives of functions of time by a dot; thus, the dots in model (6.5) denote derivatives of functions $\hat{V}(t)$ and $\hat{T}(t)$. These functions are assumed, by definition, to be continuous on the required (closed) time interval and also to be almost everywhere (continuously) differentiable on the same (but open) interval (assumptions that are important for mathematical operations on these functions). Volume and temperature do not take negative values (and zero values are meaningless), so we are only interested in the points $[\hat{V}(t), \hat{T}(t)]$ lying in the first quadrant of the "coordinates" V-T. However, a given body in general may not have the ability to occupy any point $[V, T]$ of this quadrant—this may be limited by its structure, constitution. Thus, the set of all volumes and temperatures that a body can occupy in this quadrant is called its *constitutive domain*.[6] Let's denote it $\mathcal{D}$. It

[6] Mathematically, we consider it a non-empty, continuous, open set.

thus forms the domain of the definition of our pair of independent variables. It is the equations of state that define this constitutive domain. We can now (mathematically) refine our definition of heating (6.5). The functions $\hat{\Lambda}_V$ and $\hat{C}_V$, that appear in it, defined in $\mathcal{D}$, are continuous and have continuous partial derivatives there. These functions determine the heating during an event in which the volume and temperature change.

The constitutive domain of an ideal gas is evidently the entire first quadrant of the volume-temperature coordinates. From the ideal gas equation of state (4.8) we see that for each V-T pair, i.e., for each point $[V, T]$ we can easily find the corresponding pressure from this equation, so we are not limited by the choice of the point in that quadrant. In the case of the Van der Waals equation—the Van der Waals fluid—the limitations are already obvious. Equation (5.2) states $V_m > b$, or Eq. (5.7) implies $V_{m,r} > (1/3)$. Thus, the points where these inequalities are not satisfied lie outside the constitutive domain of the Van der Waals fluid.

Part of the boundary of the constitutive domain of liquid water is shown in Fig. 6.1. It has been reconstructed from available experimental data [2–7] and it is the liquid region located above the line in the figure. Furthermore, points corresponding to several specific pressures (isobars of liquid water) are plotted in the figure, also obtained from published data [4, 8–10]. The circle marks the triple point, i.e., the

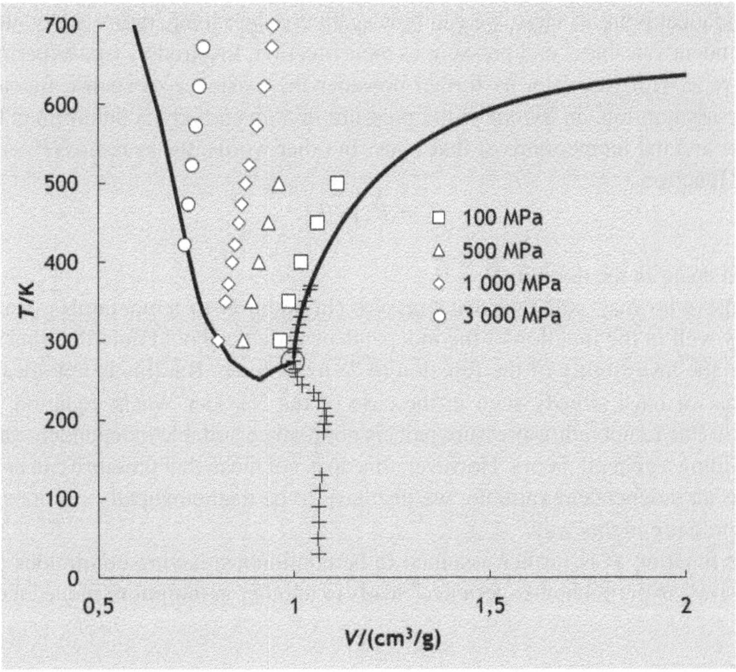

Fig. 6.1 Part of the boundary of the constitutive area of liquid water, together with several isobars; see text for explanation

point at which water exists simultaneously as liquid, vapour and ice. Finally, the points corresponding to the cooling of liquid water under atmospheric pressure to super-cooled liquid states [10], which are points extending below the shown boundary where water would otherwise exist in the solid state, are indicated by a cross in the figure. The right-hand branch of the boundary would continue on to a critical point beyond which the differences between liquid and gas become blurred.

6.3.3 Pressure as a Function

Pressure is indeed an important quantity characterizing in particular the (thermal) behaviour of gases. Despite the fact that we have hinted that it may not be as simple a quantity as it would seem at first sight, we have dealt with it in a basic way, and can thus answer the question of how to deal with pressure in our basic macroscopic model. In the experiments with gases, as we have seen, the fundamental quantity has been the pressure of a gas occupying a volume at a given temperature; there is a relationship between the three quantities just mentioned, a connection given by the nature of the gas; we have seen, for example, the equation of state of an ideal gas or the Van der Waals equation. We then see temperature and volume as adjustable parameters and gas pressure as a consequence of their adjustment. From a mathematical point of view, we can thus again consider temperature and volume as independent variables[7] and pressure as their function. Inspired by true experimental experience with gases, let us further consider the existence of such a function as a basic assumption, an axiom[8]—the pressure in a given body is determined by the volume and the temperature of that body. In other words, the pressure (P) is given by the function:

$$P = \hat{P}(V, T), \tag{6.7}$$

defined again in the domain $\mathcal{D}$.

Those who may still feel that pressure (in addition to temperature) can serve equally well in the function of the independent variable should note that this would require the invertibility of the function (6.7) for volume. But this is not always the case, as we have already seen in the case of the Van der Waals equation. Thus, pressure (the temperature-pressure pair) is not really a suitable independent variable for building a general theory. However, this does not mean that pressure can never be used as an independent variable; we just have to be mathematically cautious when using pressure in this way.

The function $\hat{P}$ is further assumed to be continuous, having continuous partial derivatives; experimental experience[9] leads to another assumption, indeed a condition:

[7] And this is what we have done so far in our macro-modelling.

[8] It was probably first formulated by Euler [11].

[9] It says that as the pressure increases, the volume (of the gas) decreases.

$$\partial\hat{P}/\partial V < 0. \tag{6.8}$$

The general Eq. (6.7) is the simplest general model describing the behaviour of gases, or liquids, briefly fluids, and may be called the constitutive (state) equation of the fluid in its simplest general form. Concrete examples are the equations of state of an ideal gas (4.8) or the Van der Waals fluid (5.1).

Heating (Q_f) can then be alternatively expressed in terms of pressure. Equation (6.7) implies that[10]

$$\dot{P} = \frac{\partial\hat{P}}{\partial V}\dot{V} + \frac{\partial\hat{P}}{\partial T}\dot{T}. \tag{6.9}$$

Since (6.8) applies, we can express the time derivative of the volume from equation:

$$\dot{V} = \frac{\dot{P} - \frac{\partial\hat{P}}{\partial T}\dot{T}}{\frac{\partial\hat{P}}{\partial V}} \tag{6.10}$$

and substitute it into (6.5):

$$Q_f = \frac{\Lambda_V}{\frac{\partial\hat{P}}{\partial V}}\dot{P} + \left(C_V - \Lambda_V\frac{\frac{\partial\hat{P}}{\partial T}}{\frac{\partial\hat{P}}{\partial V}}\right)\dot{T} \equiv \Lambda_P\dot{P} + C_P\dot{T}. \tag{6.11}$$

Here Λ_P is the *coefficient of thermal pressure change* and C_P is the *heat capacity at constant pressure*. We can easily see that the unit of C_P is again J K^{-1}, the unit of Λ_P is m^3.

6.4 Mathematical Description of Thermal Behaviour

6.4.1 What Is Meant; the Processes

What does it mean to describe the thermal behaviour of a body? In particular, to show it in the area $\mathcal{D}$, to mark a line in it, corresponding to some particular thermal behaviour, connecting the points $[V, T]$ corresponding to this behaviour. Intuitively, by behaviour we mean the behaviour of the body in time, but it can also mean that nothing happens to the body in time, it does not change, it has simply come to "rest", to equilibrium. Thus, for example, an ideal gas placed in a given volume in a thermostat at a given temperature has a constant volume and temperature and,

[10] Almost everywhere, i.e., except a finite number of time points.

according to Eq. (4.8) pressure in such an equilibrium. It then occupies a single, time-invariant point in the V-T coordinates. To describe the behaviour of a body in such an equilibrium, we need only its equation of state (constitutive), giving the relationship between volume, temperature and pressure (for the simpler bodies, to which we are limited for now). Thermal behaviours in which volume, temperature (and perhaps even pressure) change are certainly richer, more interesting. Let's take a closer look at them.

Let's call the way, how the volume and the temperature of the body change simultaneously in time, the *process* (happening in time). The thermal behaviour of a body is then understood as its behaviour during the given (thermal) process. The process can therefore be represented in coordinates already known in the constitutive domain of the body in question. Mathematically, the process, running from time t_1 to time t_2 is a piecewise continuous mapping of the interval $\langle t_1; t_2\rangle$ into the domain $\mathcal{D}$. The mapping is described by a pair of functions $\hat{V}(t)$ and $\hat{T}(t)$, which are continuous on the interval $\langle t_1; t_2\rangle$ and almost everywhere continuously differentiable in it. During the process, at time t the points $[\hat{V}(t), \hat{T}(t)]$ of the constitutive domain are occupied. These points form some *curve* in $\mathcal{D}$ which we denote by $\mathcal{P}$ (as process).

For now, we are concerned with simple thermal behaviour and its model (6.5), which captures the change in volume or temperature due to heating, and this represents the energy flow. The energy exchanged by heat during the process that takes place between time instants t_1 and t_2 is given by the integral (6.1), which for our basic model (6.5) is a *line integral* along the curve corresponding to the process $\mathcal{P}$.[11] That is, along the curve formed by the points $[\hat{V}(t), \hat{T}(t)]$:

$$Q \equiv Q(\mathcal{P}) = \int_{\mathcal{P}} (\Lambda_V \mathrm{d}V + C_V \mathrm{d}T). \tag{6.12}$$

So all the processes that correspond to the same curve $\mathcal{P}$ give the same value of $Q(\mathcal{P})$. From a formal mathematical point of view, time t is a parameter of that curve. In this sense it is unimportant—any function of t (with positive derivative) would parameterize that curve equally well. After all, time has somehow disappeared directly from Eq. (6.12). This has led classical, so-called equilibrium thermodynamics, to ignoring time, while not keeping in mind that this is only in the special case of model (6.5).

Thus, the curve $\mathcal{P}$ is generated by all such processes, which we will call *equivalent*, and which by definition assume the existence of a (piecewise smooth, increasing) transformation of the parameter—let us denote it by $\hat{\phi}$. That definition states that with some process $[\hat{V}(t), \hat{T}(t)]$ on $\langle t_1; t_2\rangle$ there is an equivalent process, denoted by a star, on $\langle t_1^*; t_2^*\rangle$, if ϕ transforms $\langle t_1; t_2\rangle$ to $\langle t_1^*; t_2^*\rangle$ in such a way that $\hat{V}(t) = \hat{V}^*(\hat{\phi}(t))$ and $\hat{T}(t) = \hat{T}^*(\hat{\phi}(t))$ for all $t \in \langle t_1; t_2\rangle$.

If we reverse the process—even if only in thought—i.e., instead of from t_1 to t_2 it goes from t_2 to t_1, we get similarly the curve $-\mathcal{P}$ corresponding to the *reversed process*, virtually to all equivalent reversed processes. Formally, mathematically, the

[11] We therefore denote the process and its corresponding curve by the same symbol.

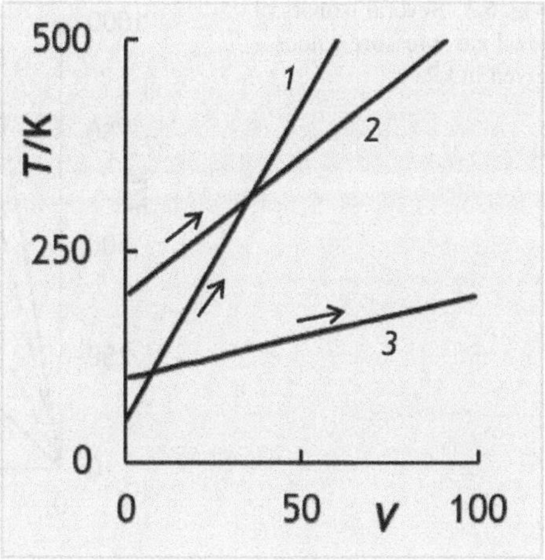

Fig. 6.2 Examples of processes of linear growth in volume and temperature over time, at different rates; the arrow shows the flow of time

reversed process is defined by a pair of functions on $\langle t_1; t_2\rangle$, which at time t have values $\hat{V}(t_1+t_2-t)$ and $\hat{T}(t_1+t_2-t)$. The basic properties of the curve integral imply the following:

$$\mathcal{Q}(-\mathcal{P}) = -\mathcal{Q}(\mathcal{P}). \tag{6.13}$$

Let's look at some simple examples of processes. For example, the linear growth of both volume and temperature over time is primitive—examples are given in Fig. 6.2. However, the independent variables of temperature and volume can take generally arbitrary values from the constitutive domain of a given body (gas). To deal with examples occupying the whole area or some part of it at once is quite elusive. It is much easier and more illustrative to look at rather narrowly defined examples. Two, even more primitive than the previous ones, present themselves naturally as processes in which one of the two independent variables is constant—that is, *isothermal* and *isochoric* processes. These are represented in the T-V coordinates simply as straight lines lying in the constitutive domain, parallel to one of the coordinate axes and having the minimum and maximum temperature or volume in the given process as end values. This representation is even apparently quite general, independent not only of the type of gas but also of the state of the body. The representation is very simple but illustrative, and the reader is warmly encouraged to sketch it himself or herself.

6.4.2 *Isobars*

Another, a little more complex example is such a combination of temperature and volume, which always gives a constant (i.e., always the same) pressure—i.e., *isobaric* process. Its representation in T-V coordinates depends also on the nature of the gas

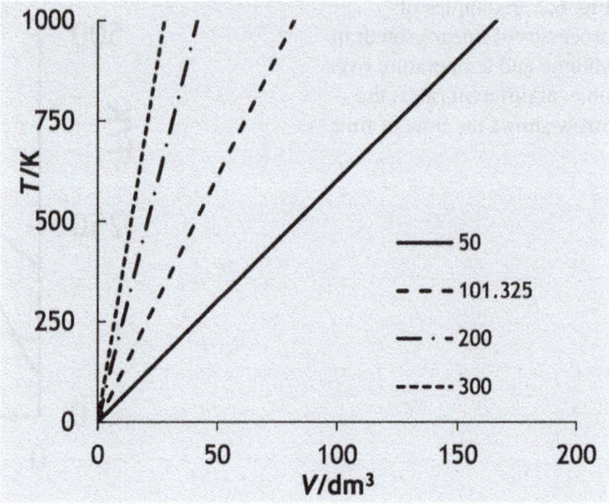

Fig. 6.3 Several isobars of ideal gas, pressure values given in kPa

(the body in general). It is simplest in the case of an ideal gas, where it is the respective sections of "rays" (lines) passing through the origin. See Fig. 6.3. In terms of time, this is, for example, a process in which both temperature and volume change linearly with time (not necessarily at the same rate).

In the case of a Van der Waals fluid, the isobars are more complex, but thus also more interesting. Several are shown in Fig. 6.4 including the one for zero (reduced)

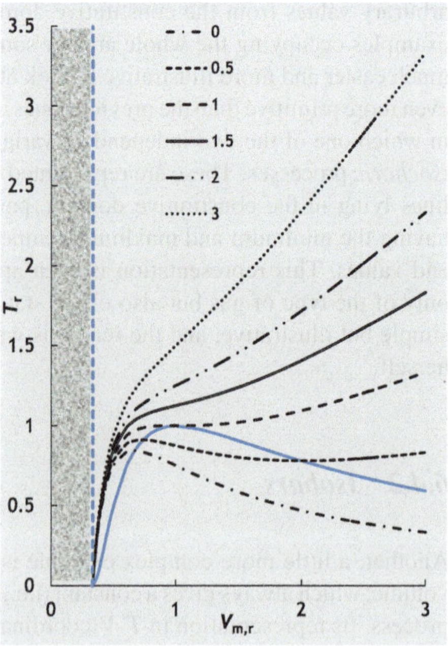

Fig. 6.4 Van der Waals fluid isobars for several values of reduced pressure; the grey part lies completely outside the constitutive domain; for further explanation see the text

pressure. However, it largely extends into the region where condition (6.8) is violated, which is the area under the $\partial \hat{P}_r/\partial V_{m,r} = 0$ curve, also marked in the figure (solid blue line). Thus, the real constitutive domain of the Van der Waals fluid will be considered as the area above the latter curve and with a volume satisfying the condition $V_{m,r} > 1/3$. A small part of it—that enclosed between the zero-pressure curve (dot-dashed in Fig. 6.4) and the blue curve $\partial \hat{P}_r/\partial V_{m,r} = 0$—includes negative pressure values, which is sometimes taken to be the region of fluid supporting tension instead of pressure. Isobars with a positive value of reduced pressure, but less than 1, form two separate branches in the constitutive domain, and here we see the isotherms that intersect them both. This means that the entire composite isobar can intersect a certain isotherm at two different values of volume. The larger value is attributed to the gas, the smaller to the liquid. Similarly, the isotherms for values of $T_r < 1$ have two separate branches in the region of validity of condition (6.8) The isobars with values of $P_r > 1$ then intersect each isotherm at only one point.

From Fig. 6.4 alone, we see that for a Van der Waals fluid, we cannot always invert the function (6.7) for volume. This is an illustrative example again of why we cannot use the pressure-temperature pair as truly general independent variables for general thermodynamic theory.

The *isobar* is represented in the coordinates of our independent variables T-V by a curve, i.e., the set of points $\hat{T}(t)$ and $\hat{V}(t)$, through which the body passes in time during the isobaric process. Along this curve the pressure is constant, deriving the equation for the pressure (6.7) along the isobar considering the functions $\check{V}(t)$ and $\check{T}(t)$ we get

$$\frac{\partial \hat{P}}{\partial V}\frac{dV}{dt} + \frac{\partial \hat{P}}{\partial T}\frac{dT}{dt} = 0. \tag{6.14}$$

The letter t here denotes time, but from a mathematical point of view it is any parameter of the function $\check{V}$ and $\check{T}$ determining the given curve (isobar). The functions representing the curve are assumed, as is usual in mathematics, to be continuously differentiable for a parameter t lying on some (finite or infinite) interval and that their derivatives dV/dt and dT/dt do not vanish simultaneously for any value of t. The last assumption, together with condition (6.8), leads to the conclusion that $dT/dt \neq 0$[12] always holds on the isobar. Thus, the function $\check{T}(t)$ is invertible and the volume can then be expressed as a function of temperature along the isobar. Moreover, this function is obviously continuously differentiable and satisfies the condition

$$\frac{dV}{dT} = -\frac{\dfrac{\partial \hat{P}}{\partial T}}{\dfrac{\partial \hat{P}}{\partial V}}. \tag{6.15}$$

Equation (6.15), on the other hand, can be viewed as a differential equation for volume (as a function of temperature). We can easily see that its solution on some

[12] Try to justify why. If you fail to do so, wait for Sect. 9.3.1.

temperature interval is an isobar valid for a body with a given pressure function $\hat{P}$. It is therefore a differential equation of an isobar, its right-hand side is a continuous function of volume and temperature, and it follows from the existential theory of solutions of differential equations that an isobar passes through each point of the constitutive area (the equation has a solution there). Although the properties of those functions are not sufficient for the uniqueness of such an isobar, this in turn follows from condition (6.8) and the theorem on implicit functions.[13] Thus, the mathematics confirms both the existence and the uniqueness of isobars.

This is not all that mathematics has to teach us about isobars. When the numerator of the right-hand side of Eq. (6.15) is zero at some point, we see that the isobar passing through that point intersects the isotherm "normally", which means that the tangent to the isobar is parallel to the temperature axis at that point and thus makes a right angle with the isotherm passing through the same point. The point at which the following applies

$$\frac{\partial \hat{P}}{\partial T} = 0, \tag{6.16}$$

will be called[14] the *piezotropic* point. At all other points in the isobar is $dV/dT \neq 0$. The observation of this paragraph can be restated as follows: along the isobar, the volume passes through the extremes only at the piezotropic points; these points can form a curve—a *piezotrope*.

From the equation of state of an ideal gas, we can easily see that for it the equality (6.16) does not hold at any point, similarly it does not have zero derivative dV/dT. Thus, the ideal gas has no piezotropic points. Also, for the Van der Waals fluid, we find that the following applies in its constitutive area, for example, $\partial \hat{P}_r/\partial T_r > 0$, so it too has no piezotropic points.

6.4.3 Water as a Challenge

Water, on the other hand, is pretty much universally known to have (at atmospheric pressure) a maximum density in the liquid state (which is why ice floats on water), and it does so at a temperature of 3.924 °C. Density is the inverse of (specific) volume, so the maximum of density is also the minimum of that volume. We do not want to get into a discussion of the behaviour of water here, let alone evaluate the models (equations of state) that have been proposed to describe it. We will only use the published findings as a brief illustration of what a general thermodynamic theory (of fluids) should deal with if it is to be truly general.

Experimental experience shows that the density maximum or volume minimum in liquid water also occurs at other pressures, and several examples of the relevant

[13] More precisely, it is unambiguous in the sense that two isobars defined on the same temperature interval are the same.

[14] Following Truesdell and Bharatha [12].

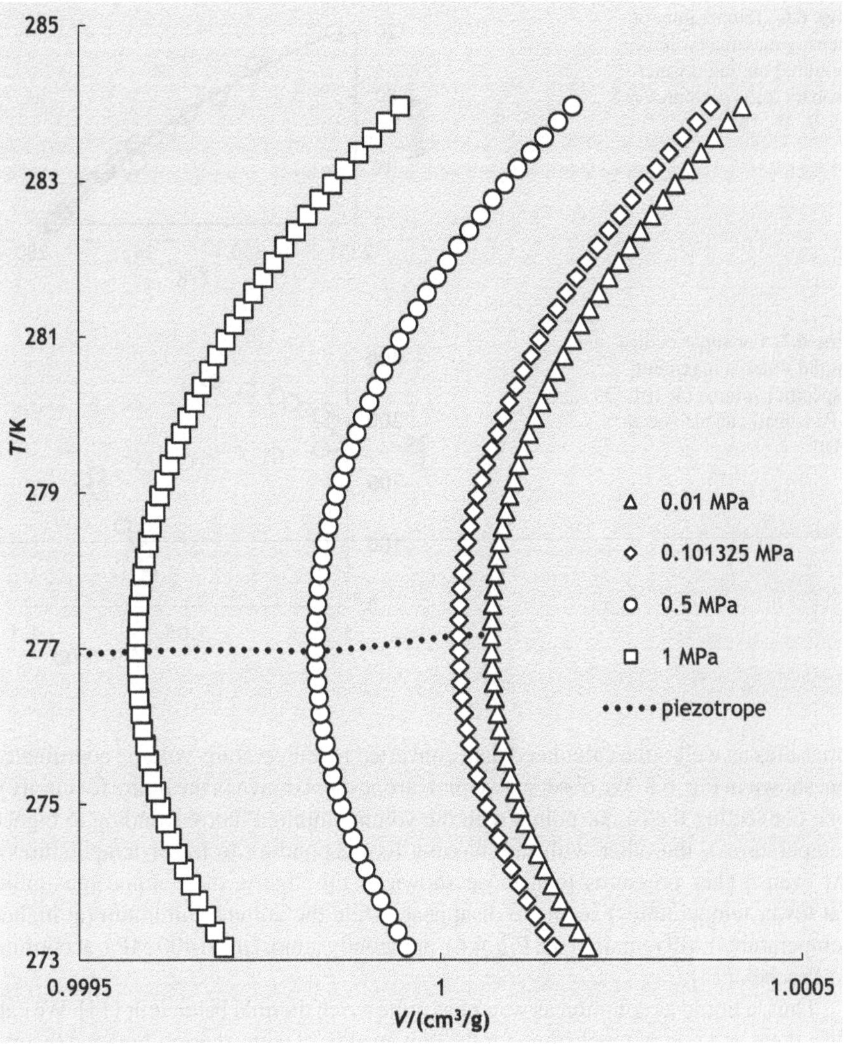

Fig. 6.5 Several isobars of liquid water with (specific) volume minima; data [6]

dependencies in our favourite temperature-volume coordinates are shown in Fig. 6.5. At even higher pressures than those shown in this figure, the extremes shift to ever lower temperature values; a summary of selected experimental data is shown in Fig. 6.6—the curve going through these data points is thus piezotrope. If we manage to super-cool the liquid water (i.e., cool it without solidifying it), we find that it even has a maximum volume—see the experimental data in Fig. 6.7. The data are for atmospheric pressure; we obviously do not have experimental data for other pressures, but model calculations show that the volume maximum occurs at other

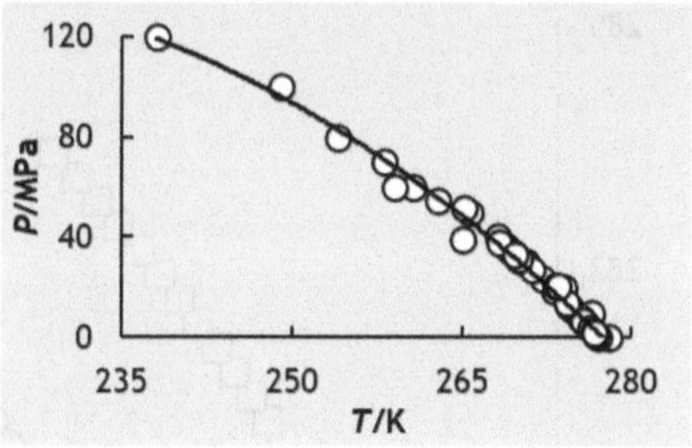

Fig. 6.6 Temperature of density maxima (volume minima) on liquid water isobars shifts with pressure [5, 7, 15, 16]

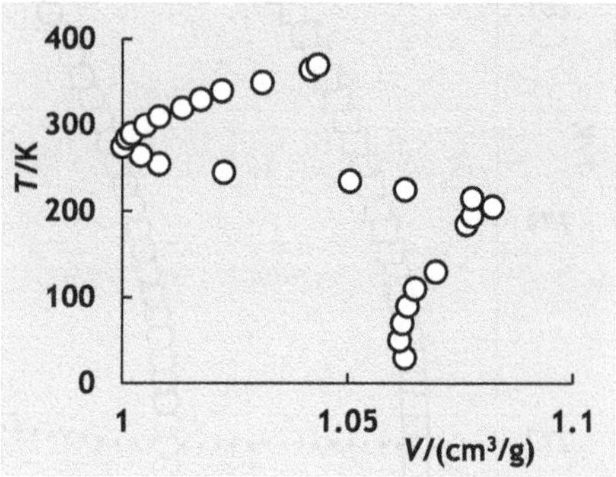

Fig. 6.7 For super-cooled liquid water, a maximum (specific) volume (at 101.325 kPa isobar) can also occur [10]

pressures as well—the calculated data converted to temperature-volume coordinates are shown in Fig. 6.8. We observe two piezotropes (not drawn in the figure for clarity), one connecting the isobar points with the volume minima (corresponding to higher temperatures), the other with the maxima (corresponding to lower temperatures). At even higher pressures than those shown in this figure, the volume maximum (at lower temperatures) seems to disappear, while the volume minimum (at higher temperatures) still remains (see Fig. 6.6), apparently almost up to 400 MPa, according to Paschek [13].

Thus, a liquid as common as water has quite a rich thermal behaviour [14]. We can take them as a sort of touchstone for the universality of some thermodynamic theory. From Fig. 6.8 we can clearly see that we cannot take the pressure-volume pair as a pair of independent variables in a truly general theory. Indeed, for a given pressure-volume value, we cannot always find a single corresponding temperature value. In other words, we cannot always invert the function $\hat{P}$, Eq. (6.7), for temperature and replace the temperature-volume pair of variables by a pressure-volume pair, as shown by the definition of the piezotropic point (6.16).

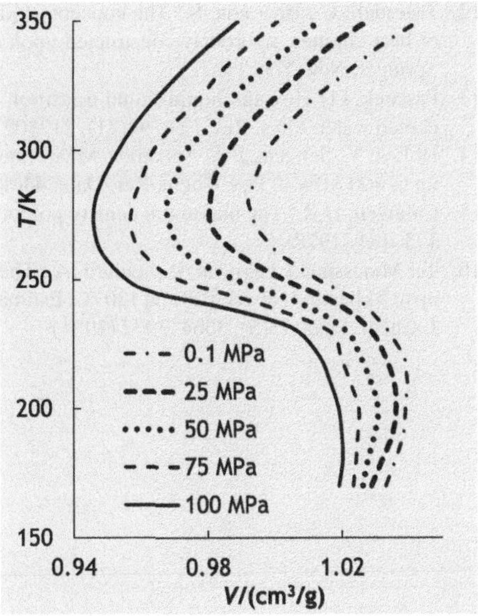

Fig. 6.8 Calculated isobars of super-cooled liquid water [13]

References

1. Truesdell, C.: The Tragicomical History of Thermodynamics 1822–1854. Springer, New York (1980)
2. Aleksandrov, A.A., Chasanšin, T.S., Larkin, D.K.: Issledovanije plotnosti obyčnoj i t'aželoj vody vblizi krivoj plavljenija. Žurnal fizičeskoj chimiji **50**(2), 394–396 (1976)
3. Asahara, Y., Murakami, M., Ohishi, Y., Hirao, N., Hirose, K.: Sound velocity measurement in liquid water up to 25 GPa and 900 K: implications for densities of water at lower mantle conditions. Earth Planet. Sci. Lett. **289**(3–4), 479–485 (2010)
4. Bollengier, O., Brown, J.M., Shaw, G.H.: Thermodynamics of pure liquid water: sound speed measurements to 700 MPa down to the freezing point, and an equation of state to 2300 MPa from 240 to 500 K. J. Chem. Phys. **151**(5), 054501 (2019)
5. Mishima, O.: Volume of supercooled water under pressure and the liquid-liquid critical point. J. Chem. Phys. **133**, 144503 (2010)
6. Linstrom, P.J., Mallard, W.G. (eds.): NIST Chemistry WebBook, NIST Standard Reference Database Number 69. National Institute of Standards and Technology, Gaithersburg. https://doi.org/10.18434/T4D303 (visited June 6, 2022)
7. Sotani, T., Arabas, J., Kubota, H., Kijima, M.: Volumetric behaviour of water under high pressure at subzero temperature. High Temp. High Press. **32**, 433 (2000)
8. Abramson, E.H., Brown, J.M.: Equation of state of water based on speeds of sound measured in the diamond-anvil cell. Geochimica et Cosmochimica Acta **68**(8), 1827–1835 (2004)
9. Belonoshko, A., Saxena, S.K.: A molecular dynamics study of the pressure-volume-temperature properties of super-critical fluids: I. H_2O. Geochimica et Cosmochimica Acta **55**(1), 381–387 (1991)
10. Mallamace, F., Branca, C., Broccio, M., Corsaro, C., Mou, C.-Y., Chen, S.-H.: The anomalous behavior of the density of water in the range 30 K $<T<$373 K. Proc. Natl. Acad. Sci. **104**(47), 18387–18391 (2007)
11. Truesdell, C.: Rational Thermodynamics, 2nd edn. Springer, New York (1984)

12. Truesdell, C., Bharatha, S.: The concepts and logic of classical thermodynamics as a theory of heat engines, rigorously constructed upon the foundation laid by S. Carnot and F. Reech. Springer, New York (1977)
13. Paschek, D.: How the liquid-liquid transition affects hydrophobic hydration in deeply supercooled water. Phys. Rev. Lett. **94**(21), 217802 (2005)
14. Holten, V., Sengers, J.V., Anisimov, M.A.: Equation of state for supercooled water at pressures up to 400 MPa. J. Phy. Chem. Ref. Data. **43**(4), 043101 (2014)
15. Caldwell, D.R.: The maximum density points of pure and saline water. Deep. Sea Res. **25**(2), 175–181 (1978)
16. Ter Minassian, L., Pruzan, P., Soulard, A.: Thermodynamic properties of water under pressure up to 5 kbar and between 28 and 120 °C. Estimations in the supercooled region down to −40 °C. J. Chem. Phys. **75**(5), 3064–3072 (1981)

Chapter 7
Heat Machines and Cycles—Internal Energy

Abstract The relevance of heat machines and cyclic processes in the analysis of thermal behaviour is addressed. The mathematical properties of cycles are presented, particularly Green's theorem. Another important quantity—work—is introduced; volume work is stressed. The First Axiom, stating the equivalence of energy transfer by heat and work in the cycle, is formulated and its historical background is given. The First Axiom naturally leads to a function of temperature and volume called internal energy. The latter term is explained as well as the associated naming of the first law of thermodynamics, which is derived from the first axiom.

7.1 Why Machines and Cycles?

After specific examples, we turn to theorizing. A special position among thermal processes is occupied by the thermal behaviour of bodies (mainly gases) in cyclic processes. Why just *cycles*? It comes from the same source as the beginnings of thermodynamics—from considerations about the functioning of heat machines, about the change of the transfer of energy by heat into the transfer by work. In fact, the behaviour of gases in particular, induced by changes in temperature, has important practical implications, which in fact gave rise to thermodynamics. If the volume of a gas changes, namely by heating, the walls of the container in which it is enclosed may move. This motion can be transferred to the motion of wheels, for example, and used to propel machines or to make vehicles moving. A typical example is a gas enclosed in a cylinder with a moving piston. This is at the heart not only of steam engines, which significantly kick-started the development of thermodynamics, but also of internal combustion engines. The piston, pushed by the expanding gas, is the only well-moving wall. The motion of the piston is converted by a set of shafts and other gears into, for example, the motion of the wheels of a car.

But no body can expand (or contract) indefinitely. Let us recall, for example, the rather small space that the piston in the cylinder of an internal combustion engine has for its motion. In order to make the motion-propulsion capabilities of thermal expansion practically usable, we must return the body to its original state (volume) after it has reached the maximum possible expansion, so that it is ready for repeated

© The Author(s), under exclusive license to Springer Nature Switzerland AG 2024
M. Pekař, *The Essentials of Thermodynamics*, Physical Chemistry in Action,
https://doi.org/10.1007/978-3-031-60321-1_7

expansion, for continuous propulsion. Heat machines (engines) thus necessarily work in a cycle. Think again of the cyclic motion of the piston in the internal combustion engine of a car. The piston cylinder can only operate up to the extreme position of the piston in which the entire cylinder (its internal volume) is "free" underneath. If the piston is to continue to move and continue to drive the vehicle, it must be returned to the opposite extreme position, or to the initial position, from which the piston can move again due to further expansion of the gas enclosed underneath. The piston therefore cycles between two extreme positions—in one motion it transfers energy to the moving vehicle, in the other "we" have to expend some energy to return the piston to the initial position, to start the next cycle. If we are to benefit from all the cycles, the energy transferred in the first step must be greater than the energy we must expend to return the piston.

7.2 A Quick Mathematical Look at Cycles

Cyclic processes have also mathematical charms. The first and quite simple fact is that the integral along a closed curve of the total differential of a function is zero. A cycle is represented in "our" temperature-volume coordinates by a closed curve—the values of temperature and pressure change in time during the corresponding cyclic process so that they lie on this curve and eventually return to the starting point; an example is shown in Fig. 7.1, where the process can take place, for example, in the direction A → B → C → D → A. Then for, e.g., the function $\hat{f}(V, T)$ the following applies:

$$\oint df \equiv \oint \left(\frac{\partial \hat{f}}{\partial V} dV + \frac{\partial \hat{f}}{\partial T} dT\right) = 0. \tag{7.1}$$

The circle through the integral symbol denotes the integral over a cycle, i.e., along a closed curve. We will see the other charm after we will discuss more thoroughly another important variable motivated by the practical use of thermal behaviour and thermal (heat) machines—work.

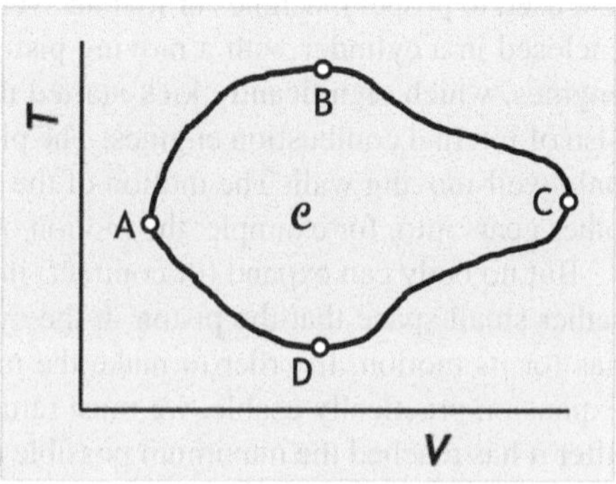

Fig. 7.1 Example of cyclic process in our preferred coordinates

7.3 Another Important Quantity—Work

Heat engines use a force "hidden" in pressure, the "pressure force", to move vehicles or machines. If a general force moves a body along a certain trajectory, we arrive at work, another quantity describing motion, which is closely related to (kinetic) energy. *Work* is determined by the product of the applied force and the trajectory travelled, or the integral of this product over the motion. From mechanics we know that it is equal to the change in kinetic energy, and work is thus another way (process) of energy exchange between bodies. If the motion is due to the action of pressure, the pressure force hidden in it, which the change of volume is related to, the work, or rather the energy exchanged by the work, is defined by these two quantities as follows:

$$W = -\int_{t_1}^{t_2} P\dot{V}\mathrm{d}t. \qquad (7.2)$$

Given the assumptions we have made about the functions $\hat{P}$ and $\hat{V}$ the above integral exists. Simple reasoning will clarify the essence of this definition. The pressure force is given by the product of the pressure and the area on which it acts: PA. The product of force and distance travelled occurs in the work, then the product of area and changing distance leads to a change in volume (over time): $PA\dot{s} = P\dot{V}$. The negative sign corresponds to the agreement that the energy transferred by the work from the body—this happens as the volume increases—is negative (somewhat popularly, the work done by the body is negative).

The work associated with a change in volume, defined by Eq. (7.2), is called *volume work*.

In our basic macroscopic model, pressure is a (composite) function of time: $P = \hat{P}(t) = \tilde{P}(\hat{V}(t), \hat{T}(t))$. The integral (7.2) determining the work is in the case of the process following from t_1 to t_2 curve $\mathcal{P}$, determined by the points $[\hat{V}(t), \hat{T}(t)], t \in \langle t_1; t_2\rangle$, the line integral—along the curve $\mathcal{P}$. So, the analogy of relation (6.13) holds for work as well:

$$W(-\mathcal{P}) = -W(\mathcal{P}). \qquad (7.3)$$

Thus, at the same time, the processes corresponding to the same curve $\mathcal{P}$ give the same value of $W(\mathcal{P})$. In other words, they exchange by work the same value of energy. But it does not mean that the processes (only) starting and ending at the same points—$[\hat{V}(t_1), \hat{T}(t_1)], [\hat{V}(t_2), \hat{T}(t_2)]$—are of this kind. These points determine only the beginning and end of the curve $\mathcal{P}$, not its shape between times t_1 and t_2. A different shape of this curve leads to a different value of the curve integral, i.e., a different value of W, as indicated in Fig. 7.2. Thus, in terms of Eq. (7.2), it depends on the particular expression of the product $P\dot{V}$ as a function of time—see the same figure.

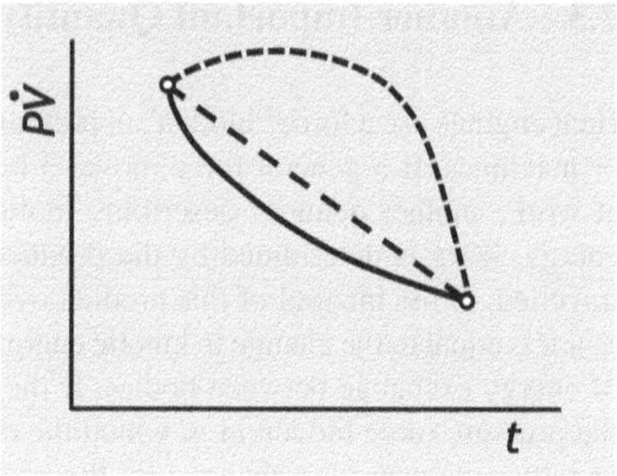

Fig. 7.2 The magnitude of work is proportional to the area under the curve connecting the initial and end points of the process

7.4 A Deeper Mathematical View of Cycles; Green's Theorem

Now back to the other promised mathematical appeal of cycles, arising from integral calculus [1]. We assume that the cycle curve $\mathcal{C}$—as in Fig. 7.1—defines a simply connected region $O_{\mathcal{C}}$ and recall that the function $\hat{P}(V, T)$ is (in the open area encompassing the cycle area) continuous and has continuous first partial derivatives. We are interested in the integral on the right-hand side of Eq. (7.2), defining the work, along a curve bounding some cycle; we see an example of a cycle again in Fig. 7.1. We can easily perform the integration by dividing the cycle curve into two parts, which can be expressed as functions of a single variable, the volume. We can certainly do this if the cycle looks something like the one just shown in Fig. 7.1, where any parallel to the temperature axis intersects it at most twice. This division is indicated in the figure by points A and C (note that these are the points with the lowest and highest values of volume in the cycle), and the two parts form curves ABC and ADC. Let us denote the function determined by the curve ABC by $\hat{T}_{\mathrm{h}}(V)$, the one determined by the curve ADC by $\hat{T}_{\mathrm{d}}(V)$. The considered integral can be expressed as the sum of two integrals over these parts:

$$\oint_{t_1}^{t_2} P\dot{V}dt = \oint PdV = \int_{ABC} PdV + \int_{CDA} PdV$$
$$= \int_{V_A}^{V_C} \hat{P}(V, \hat{T}_{\mathrm{h}}(V))dV - \int_{V_A}^{V_C} \hat{P}(V, \hat{T}_{\mathrm{d}}(V))dV. \quad (7.4)$$

However, we reach the same result by integrating the partial derivative of $\partial\hat{P}/\partial T$ over the cycle area:

$$\iint_{O_C} \frac{\partial \hat{P}}{\partial T} dV dT = \int_{V_A}^{V_C} \Big(\int_{\hat{T}_d(V)}^{\hat{T}_h(V)} \frac{\partial \hat{P}}{\partial T} dT \Big) dV$$
$$= \int_{V_A}^{V_C} [\hat{P}(V, \hat{T}_h(V)) - \hat{P}(V, \hat{T}_d(V))] dV. \tag{7.5}$$

So in the end we have equality

$$\oint \hat{P}(V, T) dV = \iint_{O_C} \frac{\partial \hat{P}(V, T)}{\partial T} dV dT. \tag{7.6}$$

Note that the whole derivation of Eq. (7.6) has been done and the equation itself is valid for a clockwise orientation of the cycle. That is, the corresponding volume-temperature pairs of values (points) in increasing time run in the A → B → C → D → A direction in Fig. 7.1. In other words, as the cycle curve runs in the direction of increasing time, the area enclosed by the cycle curve remains always on the right-hand side. For a counter-clockwise cycle, there would be a minus sign on the right side of Eq. (7.6).[1]

The relation (7.6) is a special case of the more general *Green's*[2] *theorem*, and its derivation is its simplified proof. The theorem has a more general formulation, which we will state without further proof. It starts from a simply connected domain O, which has the curve k as its boundary, and takes two functions, $\hat{P}(V, T)$ and $\hat{Q}(V, T)$ (the symbols P and Q now denote quite general functions here, not just and only pressure and heat), which are continuous in the open domain containing the domain O and also have continuous first (partial) derivatives in it. The theorem then shows the connection between the integral along the curve k and the integral over the region O ("surrounded" by k); the first integral is taken clockwise, see the preceding paragraph:

$$\oint_k (PdV + QdT) = \iint_O \Big(-\frac{\partial \hat{Q}}{\partial V} + \frac{\partial \hat{P}}{\partial T} \Big) dV dT. \tag{7.7}$$

When integrating counter-clockwise, the signs on the right side are reversed. Try to prove Eq. (7.7), or its yet unproven second part, yourself, it is very similar to the above proof of its first part of (7.6).[3] We will see later what we need Green's theorem for. Now we turn again to the application side, to heat machines.

---

[1] Try to derive for yourself.

[2] George Green (1793–1841), a self-taught English mathematician, published most of his work before 1833, when he began his studies at Cambridge University, where he later worked shortly before his death.

[3] This time you just use the division of the ABCD curve in Fig. 7.1 into the BCD and DAB parts, and use them to express the volume as a function of temperature. Adding the two proved parts gives (7.7).

7.5 Work and Cycles, the First Axiom

7.5.1 Formulation of the Axiom

Using work, we can formulate the operation of heat engines in a different way. If we give energy to an engine (generally a body) by heat, the engine can move another body, transferring energy to it by doing work, often said to do work. Simply put, heat engines (cyclically) convert some of the energy gained by heat into energy transferred by work. In practice, we are concerned with the efficiency of this conversion—ideally it would be 100%, so that all the energy absorbed by the heat could be given up by the work. In an internal combustion engine or a steam engine, we cause the gas under the piston to expand by "supplying" energy with heat; the return of the piston to its initial position can be achieved by cooling, i.e., "removing" energy with heat. Thus, in the practical operation of heat engines, we try to ensure that the energy gained by heat exceeds the energy released by heat as much as possible (and this includes "heat losses").

In terms of efficiency, the cycle is a kind of net gain in principle. If there is such a thing as "internal losses", these are zeroed out in the cycle by eventually returning the (working) body to the same state as at the beginning—any internal changes in the body are thus eliminated, as are any energy accumulations in the body. This is suggested by Eq. (7.1). For assessing the performance or efficiency of heat machines, therefore, cycles are very useful, indeed essential. This was probably first realized by Sadi Carnot, who subjected the cyclic operation of heat engines to a thoughtful analysis.

In a cycle, the body returns to its initial state at the end of the cycle. Its energy must therefore be the same at the beginning as at the end of the cycle, it does not change, see again Eq. (7.1). If it has been changed by heat or work during the cycle, these changes must therefore be reset, compensated, during the cycle; the energy gained by heat must be equal to the energy given up ("done") by work for the whole cycle. We express this simple reasoning mathematically as follows:

$$-W = JQ, \quad \text{where } J > 0 \tag{7.8}$$

Equation (7.8) is called *axiom I*; it expresses the equivalence of energy transfer by heat and work in the cycle. It holds for every cycle, and the constant of proportionality J has the same value for all fluids. In fact, we have already encountered it when mentioning Joule's experiments. Today we choose it equal one, which leads to the same units for work and heat, but for now we will leave it here in this general way for historical and consistency reasons.

7.5.2 Historical Motivations of the Axiom

We have introduced axiom I (7.8) by a simple but hopefully understandable reasoning to facilitate further interpretation. Axioms are not proved but posed as reasonable and acceptable assumptions. In fact, this axiom can be proved in our model of thermal behaviour, but it is neither simple nor short. Those interested in such a mathematically rigorous procedure are referred to the book [2]. Here we support the axiom at least again by what has proven useful in the interpretation of natural theories, ideas and models—by looking back in history, at the reasoning and practices of those who founded and first developed today's natural sciences.

We will focus our historical perspective on two men—the already familiar to us (Sadi) Carnot and Clausius.[4] On the basis of physical considerations, without any mathematical models or proofs, Carnot [3] concluded that the maximum of the "power of motion" of heat, that is, the maximum of the energy obtainable by work at the expense of the energy supplied by heat, cannot depend on the kind of body we use to do it. In a narrower sense, on the kind of gas that is enclosed in the cylinder of the heat machine. Further, Carnot argued that the heat-work conversion is universal in the cycle, by which he meant that the energy obtained in the cycle by work is equal to the energy supplied by heat multiplied by the universal function of temperature, that is, independent precisely of the kind of gas (body).

Carnot based his reasoning on the impossibility of obtaining an infinite source of motion (force) without consuming anything (such as fuel in a steam engine). If the heat-work transformation were not universal under given conditions, one could construct a pair of machines in which the less efficient one ran in reverse, and in such a "double cycle" of this machine the result would be just "excess" energy, serving as an inexhaustible source of more and more motion in further cyclic repetitions. For one machine would absorb energy by heat, the other would give out energy by heat (since the machines run in reverse), and if both energies (heat) are equal (these are the given conditions, roughly—the given fuel), they would compensate each other, unlike the energies of the exchanged work—we would have some work left from the more efficient machine precisely because of its higher efficiency. We would get this work from nothing, as the heats would be compensated.

Clausius developed Carnot's ideas further and axiom I is motivated mainly by his work [4]. He relied, among other things, on Joule's experiments, which he saw as proof that heat consists in the motion of the smallest particles of bodies. He also took his experiments as proof that if, in the course of doing work, the bodies involved are also heated, the corresponding heat is proportional to the work done. If heat is closely connected with motion, we may apply to it the general principles of mechanics as a science of motion. For example, the observation that the change in kinetic energy is proportional to the work done. All of this eventually led Clausius to the idea of the equivalence of heat and work. Yet Clausius understood it more clearly than it first

[4] Rudolf Julius Emanuel Clausius (1822–1888), a German physicist and mathematician born in present-day Poland, worked at the universities of Zurich, Würzburg and Bonn, contributing especially to the development of thermodynamics and kinetic theory of gases.

appears—he was talking about two distinct processes. About—in his words—the consumption of heat and the transfer of heat from a hotter to a colder body. Both processes can take place while work is taking place, and both corresponding amounts of heat are proportional to the work.

According to Clausius, the cycle is special in that some possible "internal work" of the body (machine), which would not be manifested by the motion of another body (the driven machine), but only by internal changes of the machine, is cancelled by returning to the initial state, and only "external work" remains, usable for example for propulsion. For Clausius, the heat-work transformation is uniform, that is, the coefficient of proportionality in the heat-work relation is not a function but a constant, the same for all bodies (fluids). It is our axiom (7.8) that is the mathematical expression of this Clausius principle.

7.6 Internal Energy and the First Law

Axiom I (7.8) in conjunction with several of the above relations, has a very interesting consequence.

Let us recall that in the cycle:

$$V(t_1) = V(t_2), \quad Q_f(t_1) = Q_f(t_2). \tag{7.9}$$

Based on the definitions of work and heat (7.2) and (6.1), we can write (7.8) as follows (in the constitutive domain $\mathcal{D}$):

$$\oint_{t_1}^{t_2} P\dot{V}\mathrm{d}t = J \oint_{t_1}^{t_2} Q_f \mathrm{d}t. \tag{7.10}$$

In the cycle, then, the following applies

$$\oint_{t_1}^{t_2} (JQ_f - P\dot{V})\mathrm{d}t = 0, \tag{7.11}$$

which immediately implies the existence of the function U such that $\oint_{t_1}^{t_2} \dot{U}\mathrm{d}t = 0$. Let us unbracket Eq. (7.11) using our model (6.5):

$$JQ_f - P\dot{V} = J\Lambda_V\dot{V} + JC_V\dot{T} - P\dot{V} = (J\Lambda_V - P)\dot{V} + JC_V\dot{T}. \tag{7.12}$$

We can write the time derivative of the function just discovered in our selected pair of independent variables as follows:

$$\dot{U} = \frac{\partial\hat{U}}{\partial V}\dot{V} + \frac{\partial\hat{U}}{\partial T}\dot{T}. \tag{7.13}$$

Comparison of (7.12) and (7.13) shows

$$\dot{U} = JQ_f - P\dot{V}, \quad \frac{\partial \hat{U}}{\partial V} = J\Lambda_V - \hat{P}, \quad \frac{\partial \hat{U}}{\partial T} = JC_V \tag{7.14}$$

(the function (6.7) was used in the second equation here). The quantity U is called the *internal energy* and is defined on the domain $\mathcal{D}$ by the above procedure.

Why internal energy? First, let us note that it is a function, in our basic model a function $\hat{U}(V, T)$. This means that for given values of volume and temperature, its value is uniquely determined. Its instantaneous time change is given by Eq. (7.13) and its change during some event between time instants t_1 and t_2 is obtained by integrating

$$\int_{t_1}^{t_2} \dot{U} dt = U(t_2) - U(t_1) \equiv U(V(t_2), T(t_2)) - U(V(t_1), T(t_1)).$$

The change is therefore also uniquely determined by the given time points, or the volume and temperature values within them. These are all characteristics that we require from energy. Equation (7.14) shows that U varies with heat and work, which are quantities describing the transfer of energy. Our heating model shows that heat transfer can be manifested by a change in temperature or in the dimensions of the body. Temperature, as kinetic theory shows, is closely related to the kinetic energy of the particles (molecules) that make up the body. Changes in dimensions are due to changes in the distances ("spacing") between these particles. The transfer of work in our considerations is also associated only with a change in the dimension (volume) of the body, the body as a whole does not move, and its total kinetic (or potential) energy does not change. Thus, it is only the energy of its molecules, which are somehow hidden in its interior, that changes. Hence, internal energy, as the energy hidden in the molecules of the body, allows the exchange of energy (by heat, work) without the body as a whole having to move. Roughly, we can say that it is the energy that a body has due to the kinetic and potential energy of its molecules and its internal atomic-molecular structure.

Equation (7.14) is usually called the *first law* of thermodynamics. It is nothing more than the law of conservation of energy, written for the special conditions that come into play in thermodynamics. That is, for the case where only internal energy is meaningful (can be exchanged, transferred), which can only be changed by heat or work.

The third equality in Eq. (7.14) shows how to get at internal energy in practice. Heat capacities are among the measurable quantities. Substituting the measured dependence of the heat capacity C_V on temperature into that third equality and then integrating (with temperature as the integrating variable) gives the internal energy (its values) as a function of temperature.

References

1. McQuarrie, D.A.: Mathematical methods for scientists and engineers. Sausalito: University Science Books (2003)
2. Truesdell, C., Bharatha, S.: The Concepts and Logic of Classical Thermodynamics as a Theory of Heat Engines, Rigorously Constructed upon the Foundation Laid by S. Carnot and F. Reech. Springer, New York (1977)
3. Carnot, S.: Reflections on the motive power of fire. Mineola: Dover (1988) (reprint of English translation)
4. Clausius, R.: Abhandlungen über die Mechanische Wärmetheorie, 1 und 2. Braunschweig: Vieweg und Sohn (1864, 1867)

Chapter 8
Further Ideas Arising from the Concept of an Ideal Gas

Abstract Two important statements on ideal gases are described: Holtzmann's assertion and Hoppe's theorem, complemented with Mayer's assertion. Then, it is shown how a new function—entropy—follows from them, employing also Green's theorem. In this way, the existence of entropy is demonstrated for an ideal gas.

8.1 Holtzmann's Assertion

At the beginning of the next section of our journey towards the generalized nature of thermodynamics, we need to look again at the history of research on the thermal behaviour of gases. In 1845, Holtzmann[1] published a statement that was and is a corollary of our axiom (7.8). We will use it as another axiom to build the foundations of the general theory of thermodynamics. Holtzmann summarized the previous experience with the behaviour of gases in heating, which is modelled by our Eq. (6.5), into the assertion that (in our modern words) the energy taken up by a gas by heat is capable of producing a mechanical effect measurable by lifting a certain mass to a height of 1 m; that is, measurable by equivalent work. Holtzmann further asserted that an ideal gas at constant temperature transfers energy by heat and then by work (and vice versa) uniformly, i.e., that the coefficient between heat and work in this transfer, the "heat-work transformation", is universal for all ideal gases, that is, it does not depend on any kind of gas. Let us express his claim illustratively in the language of our model. Heating at constant temperature is according to (6.5) as follows: $Q_f = \hat{\Lambda}_V(T, V)\dot{V}$, the work is defined by Eq. (7.2), so that Holtzmann's assertion means: $P\dot{V} = J\hat{\Lambda}_V(T, V)\dot{V}$, where J is the universal coefficient (positive constant), of course with the dimension of work per unit heat. From this equation we easily derive the announced axiom (for ideal gases):

$$J\Lambda_V = P. \tag{8.1}$$

[1] Karl Heinrich Alexander Holtzmann (1811–1865), a German physicist, active at the Polytechnic of Stuttgart from 1851, was mainly concerned with thermodynamics and mechanics.

© The Author(s), under exclusive license to Springer Nature Switzerland AG 2024
M. Pekař, *The Essentials of Thermodynamics*, Physical Chemistry in Action,
https://doi.org/10.1007/978-3-031-60321-1_8

Holtzmann's assertion of the form (8.1) actually shows what the coefficient of thermal volume change Λ_V is equal to in the case of an ideal gas. By Eq. (6.11) we have introduced the heat capacity at constant pressure, and from this definition the difference of the two heat capacities results:

$$C_P - C_V = -\Lambda_V \frac{\frac{\partial \hat{P}}{\partial T}}{\frac{\partial \hat{P}}{\partial V}}. \tag{8.2}$$

In the case of an ideal gas, we can easily express both partial derivatives from its equation of state, add up—we end up with a relation reminiscent of Holtzmann's assertion:

$$C_P - C_V = \frac{nR\Lambda_V}{P}. \tag{8.3}$$

If the difference in heat capacities of the ideal gas is constant (and positive), we get the same relation as (8.1) from Eq. (8.3). Thus, Holtzmann's assertion is equivalent to the constancy (and positivity) of the heat capacity difference for the ideal gas.

8.2 Mayer's Assertion

Comparing Eqs. (8.1) and (8.3) immediately reveals another equivalence:

$$J(C_P - C_V) = nR \qquad \Leftrightarrow \qquad J\Lambda_V = P. \tag{8.4}$$

Thus: if (8.1) applies for some positive constant J, then the first relation in (8.4) also applies. This was introduced into thermodynamics by Mayer,[2] its rather peculiar contributor. Peculiar because his contributions were rather vague, not very physically or mathematically erudite, and thus often rejected. He came up with the relationship, which we will call Mayer's assertion, by considering (in turn) the connections between energy transfer by heat and work as follows. He argued that heat is a force and can be converted into mechanical action, just for example in a steam engine. If the gas in it expands under a piston maintaining a constant pressure P, mechanical power is produced by the velocity $P\dot{V}$. At the same time, energy is supplied to the gas by heat, at the rate $C_P\dot{T}$. According to Mayer, the heat transferable by work is given by $(C_P - C_V)\dot{T}$ and is in direct proportion to that work. If the coefficient of this proportionality is denoted by J, then in the words of mathematics:

$$J(C_P - C_V)\dot{T} = P\dot{V}, \qquad P = \text{const}. \tag{8.5}$$

[2] Julius Robert von Mayer (1814–1878), a German physician interested in physics, energy conversion and conservation.

The equation of state of the ideal gas implies at constant pressure that $P\dot{V} = nR\dot{T}$, which in effect gives us Mayer's assertion in (8.4).

8.3 Hoppe's Theorem, Regnault's Measurements

An observant reader will surely ask what justifies us to use the same coefficient of proportionality J in both axioms, (7.8) and (8.1), when the former refers to the cycle and fluids in general, whereas the latter only to the isothermal behaviour of ideal gases. From the perspective of today's knowledge, we could simply answer that in both cases there is a universal equivalence between the transfer of energy by heat and work, or ultimately just a conversion between units of heat and work. In our historicizing interpretation, however, this requires somewhat more specific proof. A more stringent requirement for the heat capacities of an ideal gas, which was probably first formulated by Hoppe[3] (in the form of a mathematical theorem about the relevant functions) and supported by experiments by Regnault [1], will help us: an ideal gas has constant values of both heat capacities (not just their difference). We will use Hoppe's theorem in a slightly modified way—we will naturally require the basic thermodynamic theory under construction to hold for the ideal gas model with constant values of heat capacities.. Even for such a gas, axiom (7.8) must hold with a universal value of J, which we can easily calculate using just the ideal gas model. From the second Eq. (7.14) we express the pressure,[4] derive with respect to the temperature, reverse the order of derivatives of the internal energy, substitute from the third equality (7.14) and use the constancy of the heat capacity C_V. The result is

$$\frac{\partial \hat{P}}{\partial T} = J \frac{\partial \hat{\Lambda}_V}{\partial T} \Rightarrow J = \frac{\dfrac{\partial \hat{P}}{\partial T}}{\dfrac{\partial \hat{\Lambda}_V}{\partial T}}. \tag{8.6}$$

However, by deriving Holtzmann's Eq. (8.1), we obtain the same relation for J appearing in it, which shows the validity of using the same proportionality coefficient in both axioms discussed.

[3] Ernst Reinhold Eduard Hoppe (1816–1900), a German mathematician, professor at the University of Berlin.

[4] We warmly recommend that the esteemed reader perform the operations described in this theorem immediately.

8.4 Entropy and the Ideal Gas

The axiom built on the basis of Holtzmann's assertion has an interesting and important consequence. First, let us take a slightly different view of internal energy as a function. Equation $(7.14)_2$ implies $\hat{P} = J\Lambda_V - \partial\hat{U}/\partial V$, so that

$$\frac{\partial\hat{P}}{\partial T} = J\frac{\partial\hat{\Lambda}_V}{\partial T} - \frac{\partial}{\partial T}\frac{\partial\hat{U}}{\partial V} = J\frac{\partial\hat{\Lambda}_V}{\partial T} - \frac{\partial}{\partial V}\frac{\partial\hat{U}}{\partial T} = J\frac{\partial\hat{\Lambda}_V}{\partial T} - \frac{\partial}{\partial V}(J\hat{C}_V) \tag{8.7}$$

(in the third equation we used $(7.14)_3$). So we have a result:

$$\frac{\partial\hat{P}}{\partial T} = J\Big(\frac{\partial\hat{\Lambda}_V}{\partial T} - \frac{\partial\hat{C}_V}{\partial V}\Big). \tag{8.8}$$

The temperature derivative of the pressure in Eq. (8.8) can be expressed in the case of an ideal gas as follows:

$$\frac{\partial\hat{P}}{\partial T} = P/T, \tag{8.9}$$

as can easily be seen by expressing the derivative from the ideal gas equation of state. Axiom (8.1) can then be written more generally:

$$J\Lambda_V = T\frac{\partial\hat{P}}{\partial T} \tag{8.10}$$

and combined with Eq. (8.8):

$$\begin{aligned} 0 &= -\frac{\partial\hat{P}}{\partial T} + J\Big(\frac{\partial\hat{\Lambda}_V}{\partial T} - \frac{\partial\hat{C}_V}{\partial V}\Big) = -\frac{J\Lambda_V}{T} + J\Big(\frac{\partial\hat{\Lambda}_V}{\partial T} - \frac{\partial\hat{C}_V}{\partial V}\Big) \\ &= J\Big(-\frac{\Lambda_V}{T} + \frac{\partial\hat{\Lambda}_V}{\partial T} - \frac{\partial\hat{C}_V}{\partial V}\Big) \equiv JT\Big[\frac{\partial}{\partial T}\Big(\frac{\hat{\Lambda}_V}{T}\Big) - \frac{\partial}{\partial V}\Big(\frac{\hat{C}_V}{T}\Big)\Big] \end{aligned} \tag{8.11}$$

Because $T > 0$, $J > 0$, the square bracket in (8.11) must be zero:

$$\frac{\partial}{\partial T}\Big(\frac{\hat{\Lambda}_V}{T}\Big) - \frac{\partial}{\partial V}\Big(\frac{\hat{C}_V}{T}\Big) = 0. \tag{8.12}$$

By integrating (8.12) over a surface closed by a simple cycle (running from time t_1 to time t_2), using Green's theorem (7.7) and substituting from the model (6.5) we finally get

$$\oint_{t_1}^{t_2} \frac{Q_f}{T}\mathrm{d}t = 0. \tag{8.13}$$

By a simple cycle we mean one that (as a closed curve in the temperature-volume coordinates) runs through the boundary of a non-empty, bounded, simply connected, open set. In other words, this means that we can apply Green's theorem to it. In yet other words, processes equivalent to a simple cycle are those which do not pass through any point twice, except for the same starting and ending point, and in which $\dot{V}$ and $\dot{T}$ are not simultaneously zero at any time on any interval where they exist. Equation (8.13) implies that its integrand is a total differential (see (7.1)); in other words, there is a function of volume and temperature $\hat{S}(V, T)$, for which $Q_f = T\dot{S}$. This function is called *entropy*. We can rewrite the latter relation using the total differential entropy as follows:

$$Q_f = T\frac{\partial \hat{S}}{\partial V}\dot{V} + T\frac{\partial \hat{S}}{\partial T}\dot{T}. \tag{8.14}$$

By comparison with our heating model (6.5), we find the relationships between the coefficients of this model and the partial derivatives of the entropy:

$$\Lambda_V = T\frac{\partial \hat{S}}{\partial V}, \qquad C_V = T\frac{\partial \hat{S}}{\partial T}. \tag{8.15}$$

From the last equation we can see that

$$\frac{\partial}{\partial T}\left(\frac{\hat{\Lambda}_V}{T}\right) = \frac{\partial^2 \hat{S}}{\partial V \partial T}, \qquad \frac{\partial}{\partial V}\left(\frac{\hat{C}_V}{T}\right) = \frac{\partial^2 \hat{S}}{\partial T \partial V}. \tag{8.16}$$

Equalities (8.16), when considering the interchangeability of the partial derivatives of entropy, lead back to Eq. (8.12).

Entropy was introduced in 1865 by Clausius from the Greek term ἡ τροπὴ, meaning conversion or transformation [2], presumably because it appeared to him when considering the "conversion" of heat into work and vice versa, i.e., when considering the conversion of the transfer of energy by heat into the transfer of work. Clausius himself justifies his choice by the similarity of the words entropy and energy, which, according to him, is meant to emphasize the close affinity of their physical meaning.[5]

[5] Jaynes [3] argues that the name entropy is formed from τροπὴ, meaning something like turn, turning, and the prefix εν with the result meaning to turn the head aside, which is said to express nicely the one-sided character of entropy. We will reveal this later, in Chap. 11.

References

1. Truesdell, C.: The Tragicomical History of Thermodynamics 1822–1854. Springer, New York (1980)
2. Clausius, R.: Abhandlungen über die Mechanische Wärmetheorie, 1 und 2. Braunschweig: Vieweg und Sohn, 1864 und 1867
3. Jaynes, E.T.: The minimum entropy production principle. Annu. Rev. Phys. Chem. **31**, 579–601 (1980)

Chapter 9
Entropy in General

Abstract Entropy, introduced in Chap. 8, is generalized here for all fluids. The generalization operates on new concepts—ordinary and thermodynamic points. The corresponding proof is rather complex, and an analysis of adiabatic processes precedes it. Adiabats of an ideal gas, a Van der Waals fluid, and water are examined. Then, a special type of cyclic processes, the Carnot cycle, and two additional related axioms are described. Finally, a general proof is developed in detail. Further, it is shown how entropy can alternatively be obtained from the integration factor of the basic heating model. It is also shown that the existence of the integration factor limits efficiency in the cycle: in a cyclic process, a body cannot absorb heat without also emitting heat. The chapter closes with a contemplation on the meaning of entropy.

9.1 Generalization in a Nutshell

Thus we arrive at entropy, an important thermodynamic quantity, specific to thermodynamics, but so far only in the case of an ideal gas (and of course within the framework of the heating model (6.5)). How to generalize the procedure? This is suggested by relation (8.10), which looks more general than its ideal gas equivalent and source (8.9). It would suit us in the form

$$\frac{\partial \hat{P}/\partial T}{\Lambda_V}, \tag{9.1}$$

where we would declare that expression (9.1) has the same (and nonnegative) value for all fluids at constant temperature. Of course, in doing so, we consider only those points at which $\Lambda_V \neq 0$; we will call such points of the constitutive domain *ordinary points*. Let those who would be content with this statement as another axiom take it as such. Those who would need a justification, a proof, will soon find it in Sect. 9.5 (which the former can therefore skip). Here and now, let us take the statement as valid. If it is true for all fluids, it is also true for the ideal gas, and it is the latter that allows us to find that common value of the expression (9.1) under isothermal conditions, even in the whole meaningful (first) quadrant *V*-*T*, because it is the constitutive domain of

© The Author(s), under exclusive license to Springer Nature Switzerland AG 2024
M. Pekař, *The Essentials of Thermodynamics*, Physical Chemistry in Action,
https://doi.org/10.1007/978-3-031-60321-1_9

the ideal gas. This common value is, according to Holtzmann's statement (8.1) the following value valid for all fluids (at all ordinary points):

$$\frac{\partial \hat{P}/\partial T}{\Lambda_V} = J/T. \tag{9.2}$$

Since the ideal gas has such a large constitutive domain, (9.2) applies at any temperature.

To further generalize to other points, we take advantage of the (assumed) continuity of the relevant functions and introduce the notion of a thermodynamic point. It is such a point of the constitutive domain, any neighbourhood of which contains an ordinary point. The functions $\partial \hat{P}/\partial T$ and $\hat{\Lambda}_V$ are continuous, so at any thermodynamic point of any fluid the following applies

$$T\frac{\partial \hat{P}}{\partial T} = J\Lambda_V. \tag{9.3}$$

This brings us back to relation (8.10), but this time not restricted to ideal gases. However, it is valid only for thermodynamic points, which is sufficient for practical use. Thus, the procedure could be the same as above for ideal gases, but for the sake of variety and practice we will show a slightly different derivation of relation (8.12).

We substitute that universal value of J from Eq. (9.3) into Eq. (8.7):

$$\frac{\partial \hat{P}}{\partial T} = \frac{T}{\Lambda_V}\frac{\partial \hat{P}}{\partial T}\left(\frac{\partial \hat{\Lambda}_V}{\partial T} - \frac{\partial \hat{C}_V}{\partial V}\right),$$

which implies

$$1 = \frac{T}{\Lambda_V}\left(\frac{\partial \hat{\Lambda}_V}{\partial T} - \frac{\partial \hat{C}_V}{\partial V}\right). \tag{9.4}$$

Equation (9.4) can be further modified

$$\frac{\Lambda_V}{T} = \frac{\partial \hat{\Lambda}_V}{\partial T} - \frac{\partial \hat{C}_V}{\partial V},$$

$$0 = \frac{\partial \hat{\Lambda}_V}{\partial T} - \frac{\partial \hat{C}_V}{\partial V} - \frac{\Lambda_V}{T} \Rightarrow \frac{1}{T}\frac{\partial \hat{\Lambda}_V}{\partial T} - \frac{1}{T}\frac{\partial \hat{C}_V}{\partial V} - \frac{\Lambda_V}{T^2} = 0,$$

$$\frac{\partial}{\partial T}\left(\frac{\hat{\Lambda}_V}{T}\right) - \frac{\partial}{\partial V}\left(\frac{\hat{C}_V}{V}\right) = 0,$$

$$\frac{\partial}{\partial T}\left(\frac{\hat{\Lambda}_V}{T}\right) = \frac{\partial}{\partial V}\left(\frac{\hat{C}_V}{T}\right). \tag{9.5}$$

So we again derived Eq. (8.12). The just discussed generalization procedure again implies an integration of (8.13), but this time over a simply connected and open domain determined as a subset of the constitutive domain containing only thermodynamic points. The entropy is thus generally derived only on this subset, but as a bit above we can conclude that this is sufficient for practical use. In this context, it is worth pointing out that the internal energy derived around Eqs. (7.13)–(7.14) is derived there for the whole constitutive domain.

Let us point out an important practical consequence of relation (9.3). It allows one to determine the coefficient of the thermal volume change Λ_V from experimental data, specifically from function (6.7). This is a constitutive function, in other words a constitutive or state equation. The partial derivative of the pressure that occurs in that relationship is called the *thermal pressure coefficient*.

And one more practical view. Using relation (9.3) or (8.10) in the second Eq. (7.14) we obtain an expression for the partial derivative of the internal energy with respect to the volume, which contains only measurable quantities or dependencies—pressure, temperature, change of pressure with temperature. Integrating then with the help of experimental data, we obtain the internal energy (its values) as a function of volume. Along with a similar integration mentioned at the end of Sect. 7.6 we thus obtain a complete (experimental) characterization of internal energy as a function of temperature and volume. The relevant data are of course of a constitutive nature, so we need to add that this is a characterization of the internal energy of some particular body to which the data refer.

9.2 Thermal Behaviour in Cyclic Processes Again

Before we go into the above-promised proof, let us talk about certain special cycles and parts of them. Firstly, we will need one of them in this very proof, and secondly, they are an illustration of thermal behaviour in the temperature-volume coordinates and the associated heat-work "transformation".

We have already emphasized the practical importance of cycles—for illustrating the function of thermal machines. Throughout the cycle, axiom (7.8), holds, roughly speaking, hundred-percent equivalence of total heat-total work. However, the return of the machine (engine, piston) to the beginning of a new cycle requires some energy—if we were to return the body along the opposite but otherwise same path as the expansion, for example along the same isotherm, we would have to invest the same energy as we had previously gained through work during the expansion. The net energy "gain" in such a cycle would be zero. So we have to go back in some more efficient way.

Why did we just mention the isotherm as the expansion part of the cycle? Let us quickly explain. Thus, in particular, the heating model (6.5) shows that in the isothermal case all the energy supplied by heat is attributed to the volume change, which in turn is closely related to the (volume) work done. Furthermore. The internal energy just discovered, tells us by the first Eq. (7.14) that if the internal energy does not change (in time), then in the case of volumetric work, Eq. (7.8) also holds, even in the non-cyclic process. Then all the "delivered" heat can be recovered "as work". The second Eq. (7.14) combined with equation (axiom) (8.1) says that the internal energy of an ideal gas does not depend on the volume, in other words, it is a function of temperature only. In an isothermal process, for example, expansion, the internal energy of the ideal gas does not change. And isothermal expansion is, at a rough glance, part of the cycle of a steam engine (expansion of water vapour at the boiling point of water) and of an internal combustion engine (expansion of the exhaust gases in a tempered engine).

So if isothermal expansion, albeit of an ideal gas, is so efficient in our simple model, isothermal re-entry should be similarly efficient in the case of the cycle, but obviously at a different temperature. Let the esteemed reader deduce for him-self/herself, with the help of the equations given so far, that the volumetric work is greater in absolute value at a higher temperature for an ideal gas. It's quite simple. But how does one move to the second, "return" isotherm? We've shown that they don't intersect. Well then, move along some suitable line, connecting our end point in the *T*-*V* coordinates on the original isotherm with the initial point on the return isotherm and thus representing the suitable process in these coordinates. There may be many such line-processes and each one in general may have some energy requirements. However, there is an interesting one–adiabatic.[1] In terms of energy transfer by heat it is in fact "free", it does not need any heat.

9.3 Adiabatic Processes

9.3.1 Definition and Mathematical Description

Adiabatic process is such a process, in which the heating is zero; more precisely $\hat{Q}_f(t) = 0$ in all times t, in which the function $\hat{Q}_f$ is defined. Of course, both temperature and volume change, but within our model (6.5) in such a way that a change in one of these variables is "compensated" by a change in the other[2]:

$$\hat{\Lambda}_V(T, V)\dot{V} = -\hat{C}_V(T, V)\dot{T}. \tag{9.6}$$

Thus, according to model (6.5) in the adiabatic process, the temperature change goes "hand in hand" with the volume change, i.e., still with the possibility of doing work.

The line determining the course of the adiabatic process in our temperature-volume coordinates is called the *adiabatic curve* or *adiabat*. An adiabat can be described as a curve (in the positive quadrant *T*-*V*) by a parameter, which is practically time, but in general mathematical terms it is simply some real parameter s. This is determined by the functions $V = \check{V}(s)$, $T = \check{T}(s)$ and to be an adiabat it must be

[1] From the Greek α, no/not, and $\delta\iota\alpha\beta\alpha\acute{\iota}\nu\epsilon\iota\nu$ to pass through.

[2] Again, equality in mathematical terminology applies almost everywhere.

$$\Lambda_V \frac{\mathrm{d}\check{V}}{\mathrm{d}s} + C_V \frac{\mathrm{d}\check{T}}{\mathrm{d}s} = 0, \tag{9.7}$$

while for no value of s, $d\check{V}/ds$ and $d\check{T}/ds$ are simultaneously zero.

We know that $C_V > 0$ is given, so $d\check{V}/ds$ can never be zero, because then (9.7) could only be satisfied for $d\check{T}/ds = 0$, and this is ruled out, as we have just established. The result $d\check{V}/ds \neq 0$ implies that the function $\check{V}(s)$ is invertible, and hence we can transform the function $\check{T}(s)$ to the function $\hat{T}(V)$. Its derivative is easily found from (9.7):

$$\frac{\mathrm{d}\hat{T}}{\mathrm{d}V} = -\frac{\Lambda_V}{C_V}. \tag{9.8}$$

Since $\hat{\Lambda}_V$, $\hat{C}_V$ are continuous as functions (they also have continuous partial derivatives), the function $\hat{T}(V)$ is twice continuously differentiable (along the adiabat, we mean here). Conversely, one can easily see that every solution of the (differential) Eq. (9.8) is an adiabatic one for a body with given Λ_V a C_V. Therefore, Eq. (9.8) is the differential equation of the adiabatic curve. Since $\hat{\Lambda}_V$ and $\hat{C}_V$ have continuous partial derivatives (in $\mathcal{D}$), Cauchy's existence theorem guarantees the existence and uniqueness of the adiabatic at every point in $\mathcal{D}$ of a given body. Let us emphasize that the quantities Λ_V and C_V are given by the body, they are its characteristics, determined by the constitutive equations. Therefore, the adiabat of one body need not be the adiabat of another body.

At an ordinary point, $\Lambda_V \neq 0$ by definition. Quite analogically, we can derive that at ordinary points the function $\check{T}(s)$ is invertible and thus the function $\check{V}(s)$ can be changed to the function $\hat{V}(T)$. Since the function $\hat{\Lambda}_V$ is continuous, every ordinary point has a certain (sufficiently small) neighbourhood in which Λ_V has the same (single) sign, i.e., either $\Lambda_V > 0$, or $\Lambda_V < 0$ applies there. Since the sign of Λ_V determines the sign of the adiabat slope by (9.8), each point in this neighbourhood can be obtained as the intersection of one adiabat with one isotherm, which is the sort of thing we need to build a cycle. As in the case of (9.8) we can express the adiabats for ordinary points by the differential equation

$$\frac{\mathrm{d}\hat{V}}{\mathrm{d}T} = -\frac{C_V}{\Lambda_V}. \tag{9.9}$$

and apply the existence theorem similarly as above.

9.3.2 *Ideal Gas Adiabats*

Ideal gas is also an ideally simple example of how to derive a more specific adiabatic condition for some type of gas. We need to substitute something more specific into the (definition) Eq. (9.6) for its two coefficients. Relations (8.2) and (8.3). will help us. Let us first denote the ratio of the two heat capacities by γ:

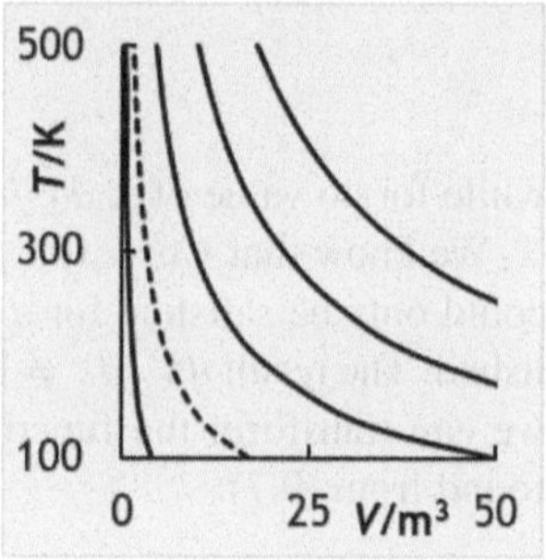

Fig. 9.1 Examples of ideal gas adiabats

$$\gamma = \frac{C_P}{C_V}. \tag{9.10}$$

Dividing Eq. (8.3) by the heat capacity at constant volume (not zero, cf. (6.6)), substituting for the pressure from the ideal gas equation of state and a simple modification gives

$$\frac{V\Lambda_V}{C_V} = (\gamma - 1)T. \tag{9.11}$$

This is already concrete constitutive information for the ideal gas. Substituting from Eq. (9.11) into Eq. (9.6) divided by C_V and then integrating, we obtain

$$TV^{\gamma-1} = \text{const.}, \tag{9.12}$$

which is one of the *Poisson's*[3] *equations* describing the adiabatic process in an ideal gas. Examples of ideal gas adiabatic curves are shown in Fig. 9.1.

9.3.3 Adiabates of Van der Waals Fluid

For the Van der Waals fluid, the relation (8.2) no longer gives us anything very nice to describe adiabaticity. But we will use the relation (9.3). From the Van der Waals equation of state (5.1) we have

$$\frac{\partial \hat{P}}{\partial T} = \frac{nR}{V - b}, \tag{9.13}$$

so that

$$J\Lambda_V = \frac{TnR}{V - b}. \tag{9.14}$$

[3] Siméon Denis Poisson (1781–1840), a French mathematician interested in mathematical applications in various branches of physics, professor at École Polytechnique and the University of Paris.

Comparing again with Eq. (5.1) we get

$$J\Lambda_V = \hat{P} + \frac{a}{V^2}. \tag{9.15}$$

So we have two equations for determining the coefficient of thermal change of the volume of a Van der Waals fluid from (measurable) values of temperature, volume or pressure. Substituting the first of these into the adiabatic form of our thermal model (9.6), we obtain the equation

$$\frac{TnR}{J(V-b)}\dot{V} + C_V\dot{V} = 0 \tag{9.16}$$

and by its integration the condition for the adiabatic process in the Van der Waals fluid:

$$T(V-b)^{nR/JC_V} = \text{const.} \tag{9.17}$$

We can convert it into a dimensionless form with reduced quantities, see the derivation of Eq. (5.7). That is, dividing by $8a/27nRb$ and by $(3b)^{nR/JC_V}$; the result is

$$T_r\left(V_{m,r} - \tfrac{1}{3}\right)^{nR/JC_V} = \text{const.} \tag{9.18}$$

Several adiabatic curves calculated from the (reduced) Van der Waals equation can be seen in Fig. 9.2.

9.3.4 Water Adiabats

For water, unfortunately, we do not have a simple enough equation of state that would allow us to easily integrate the differential adiabatic equation (9.8). This is a pity, because it is in a substance with a piezotrope that the adiabatic progression would be interesting. We shall at least attempt a rough qualitative estimate, since we are not concerned here with an analysis of the thermodynamics of water, but with an illustrative example. The data for the coefficient of thermal volume change (Λ_V) are apparently not available at all. From Eq. (9.3) we know that the temperature-pressure coefficient would be useful to find it. There is not much data for this in the case of water either, the probably isolated exception being the work [1], but that describes measurements quite far from the piezotropic points. So we shall resort to calculations using the US National Institute of Standards and Technology database [2], which uses the state formulation developed for water by IAPWS. We will restrict ourselves to the vicinity of piezotropic points that are not too far (temperature-wise) from the best-known maximum of water density. At the piezotropic point the pressure has a minimum, an example is shown in Fig. 9.3. To the left of it (lower temperature), the pressure is a decreasing function of temperature and the Λ_V coefficient is therefore

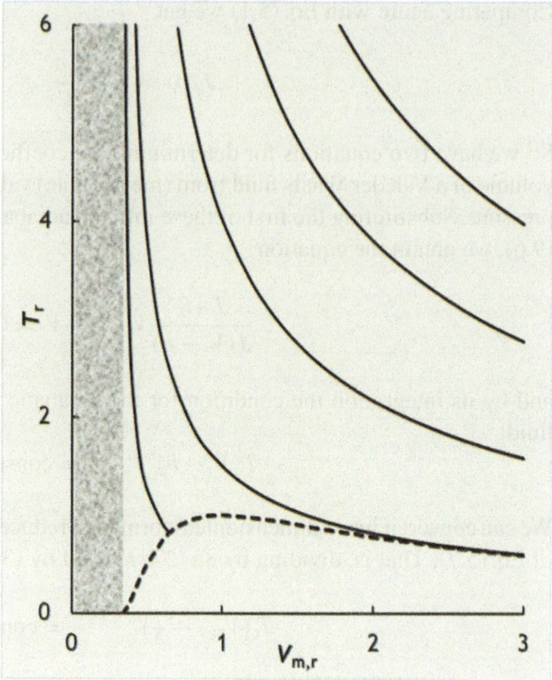

Fig. 9.2 Examples of adiabats (solid lines) of Van der Waals fluid (in its constitutive domain)

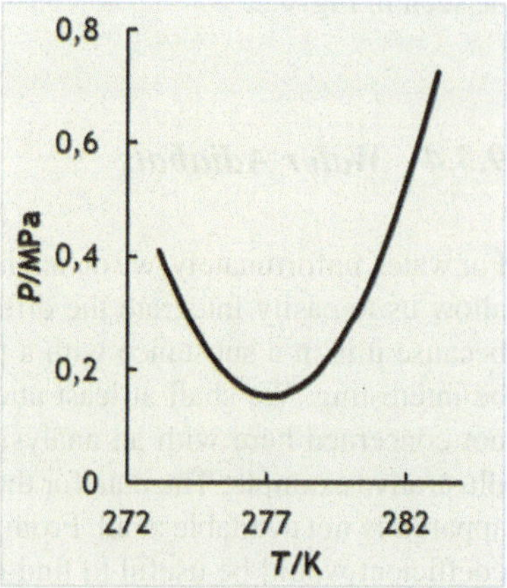

Fig. 9.3 The most famous piezotropic point of liquid water (minimum) and its surroundings

negative according to Eq. (9.3). However, according to Eq. (9.8) and relation (6.6), this implies an adiabatic pressure increasing in (our preferred) temperature-volume coordinates. To the right (higher temperature) of that point, the opposite is true. The piezotrope thus divides the constitutive domain into a part with a positive and a part with a negative Λ_V coefficient. And on the piezotrope, the adiabat has a minimum.

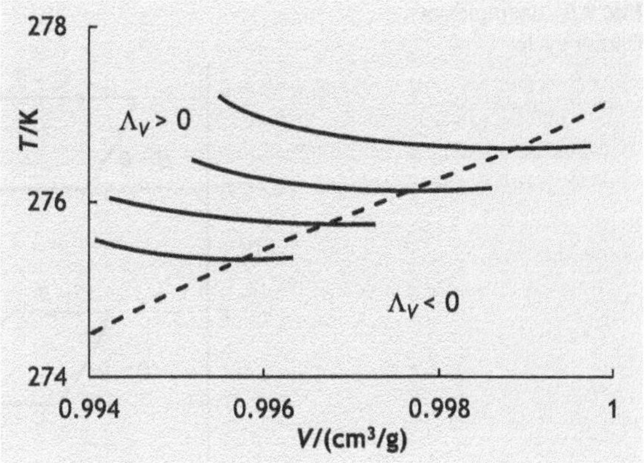

Fig. 9.4 Sketch of several probable liquid water adiabatic curves; the dashed line indicates the piezotrope

We neglect the variation of the heat capacity C_V of water with temperature and volume in the modelled, narrow region of these two quantities. From the shape of the dependence of water pressure on temperature and the related expression (9.3), in which the partial derivative is multiplied by temperature, it can be roughly inferred that the decreasing part of the water adiabatic will be more curved than the increasing part, and moreover more curved in the direction of temperature increase. The pressure minimum on the dependence is relatively flat, so the minimum on the water adiabatic will be flat as well. From these considerations comes a rough sketch of the shape of the water adiabats in Fig. 9.4.

9.4 Carnot Cycles; Other Axioms

Let us recall that we have begun to construct a model cycle composed of the isothermal and adiabatic processes of an (ideal) gas, generally some fluid. Sadi Carnot, on the basis of physical considerations, without any mathematical models or proofs, proposed just such a simple cycle, which we will call the *Carnot cycle*. The process that generates this cycle will then be called *Carnot's process*. The characteristic feature of Carnot's process is that in the exchange of energy by heat it operates only with two (different) temperatures—energy is supplied to the body by heat at a constant temperature, and it is transferred from the body at a constant temperature by heat, this time at a lower temperature. Examples of such cycles are shown in Fig. 9.5. According to Carnot, a cycle generated in this way is capable of maximum mechanical performance for a given maximum and minimum (upper and lower) operating temperatures that occur in the cycle. And it is the maximum work (energy gained in the cycle of work) that is obviously meaningful when we want to compare the efficiency of thermal machines.

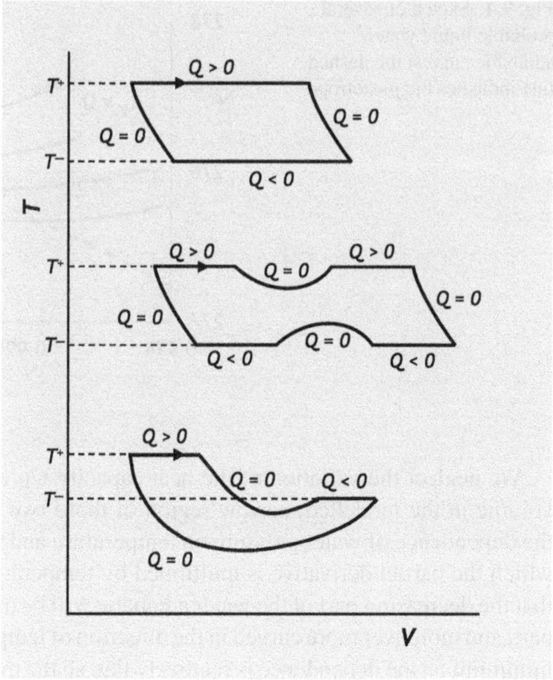

Fig. 9.5 Examples of Carnot cycle

Carnot himself writes [3, 4]: "The effects of heat in motion are independent of the substance which mediates them; their magnitude is determined only by the temperatures of the bodies between which the heat is ultimately transferred". The universality of our constant J. also resonates in his statement. But in Carnot's time the concept of an ideal gas had not yet been introduced.

A mathematically more precise definition of Carnot process says that if there are constant temperatures T^+ and T^-, for a cyclic process, $T^+ > T^-$ and

$$\begin{aligned} T &= T^+ \text{ on } \mathcal{T}^+, \\ T &= T^- \text{ on } \mathcal{T}^- \end{aligned} \tag{9.19}$$

(time intervals $\mathcal{T}^+$ and $\mathcal{T}^-$ are not empty), then this process is called Carnot's process. If the Carnot cycle further satisfies also the following three conditions: 1. it is simple, 2. its interior lies in $\mathcal{D}$, 3. Λ_V has a single sign in both the cycle and its interior, we call it an *ordinary Carnot cycle*.

There are two other axioms associated with the (ordinary) Carnot cycle, which have their source in the reasoning of both S. Carnot and Clausius. The *axiom*, which we denote by *II*, states that in ordinary Carnot cycles, $-W > 0$, i.e., there is a net gain of energy transferable by work. This was one of the reasons why Carnot designed the cycle as he did, as we explained above.

Axiom III then concerns two different fluid bodies performing (each its own) ordinary Carnot cycle. The axiom states that if the cycles of both bodies have the same operating temperatures and the same amount of energy absorbed by heat, the work (i.e., energy exchanged by work) in them is the same. This is because both cycles consume "the same amount of fuel", i.e., the energy absorbed by the heat, and operate between the same temperatures. However, they may differ (only) in the positions of the adiabats between the two isotherms.

Axiom III actually expresses Carnot's reasoning above about the impossibility of obtaining an infinite source of energy (work). According to Carnot, the operating temperatures and the energy absorbed by heat are the basic factors determining the efficiency of the (Carnot) cycle, i.e., the energy recoverable in the cycle by work on account of the energy absorbed by heat. To better understand the axiom, let us put Carnot's reasoning into the language of the mathematics of one example. Let us have two bodies, T1 and T2, each running their ordinary Carnot cycle ($\mathcal{C}$) between the same temperatures, in which they absorb the same heat. The cycle of one body, say T2, runs in reverse. Let us first look at the energies absorbed by the heat; they are assumed to be the same for both cycles: $Q^+_{\mathrm{T1}}(\mathcal{C}_1) = Q^+_{\mathrm{T2}}(\mathcal{C}_2)$. The body T2, however, runs its cycle in reverse. The energy exchanged by the heat is given by the line integral, see (6.1). The properties of line integrals, see (6.13), imply that $Q^+(-\mathcal{P}) = Q^-(\mathcal{P})$ and $Q^-(-\mathcal{P}) = Q^+(\mathcal{P})$; hence $Q^+_{\mathrm{T1}}(\mathcal{C}_1) = Q^-_{\mathrm{T2}}(-\mathcal{C}_2)$. Now the energy exchanged by the work; this is also given by the curve integral and the relation (7.3) applies to it. It then follows for our two-machine system

$$W_{\mathrm{T1}}(\mathcal{C}_1) - W_{\mathrm{T2}}(\mathcal{C}_2) = W_{\mathrm{T1}}(\mathcal{C}_1) + W_{\mathrm{T2}}(-\mathcal{C}_2).$$

If the left side of this equation were not zero, our system would be supplying energy by doing work (the right side of the equation) out of nothing, because the only sources the system would have—the energy (energies) exchanged by heat—would cancel each other out, as we showed a few lines above. This, according to Carnot, directly contradicts sound physics, certainly at least in human experience to date, hence axiom III. It can also be expressed very briefly by stating that the work in a (ordinary) Carnot cycle with given operating temperatures and given absorbed heat does not depend on the position (location) of the cycle between these isotherms, universally, for any fluid.

The distinction between the energy delivered (absorbed) to a body (cylinder with gas and piston) by heat and the energy delivered (emitted) by a body by heat[4] is a fundamental insight both for the analysis of the efficiency of thermal machines and for the development of thermodynamic theory.

[4] This distinction was actually introduced by Sadi Carnot.

9.5 Deferred Proof for the Generalization of Entropy

9.5.1 Work in Carnot Cycles

We now turn to the deferred proof that was promised in Sect. 9.1, above Eq. (9.2). The key point we now want to explore is actually the partial derivative of pressure with temperature. For this we have the thermal behaviour of a body in a cycle, which simplifies things quite a bit by its return to the initial state. And a few axioms (well supported by experience).

Let us realize that the derivative occurs in the expression giving the energy transferred by the work in the cycle ($\mathcal{C}$); in short, the work in a (simple) cycle is given by (7.6)

$$-W(\mathcal{C}) = \int\int_{\mathcal{A}} \frac{\partial \hat{P}}{\partial T} \mathrm{d}V \mathrm{d}T, \tag{9.20}$$

i.e., by Green's theorem applied to the definition of (volume) work (7.2) ($\mathcal{A}$ is the area enclosed by the cycle curve, the area in Green's theorem). For an ordinary Carnot cycle, we can write this expression in terms of a double integral, as we know from the basic calculus:

$$-W(\mathcal{C}) = \int_{T^-}^{T^+} \mathrm{d}T \int_{\hat{\phi}_1(T)}^{\hat{\phi}_2(T)} \frac{\partial \hat{P}}{\partial T} \mathrm{d}V. \tag{9.21}$$

We recall that at the adiabat the volume is expressible as a function of temperature, $V = \hat{\phi}(T)$; in Eq. (9.21) 1 and 2 number the two adiabats of the Carnot cycle. Integration is the reversed operation to derivation; what we integrate can actually be seen as the derivative of some function that we obtain by integration.[5] In expression (9.21) we integrate the second integral with respect to the temperature, so we can refer to it as $\hat{g}'(T)$:

$$\hat{g}'(T) \equiv \int_{\hat{\phi}_1(T)}^{\hat{\phi}_2(T)} \frac{\partial \hat{P}}{\partial T} \mathrm{d}V. \tag{9.22}$$

We can then write the work in (9.21) simply as the difference

$$-W(\mathcal{C}) = g(T^+) - g(T^-). \tag{9.23}$$

[5] For example, we can write $\int x^2 \mathrm{d}x$ as $\int \hat{f}' \mathrm{d}x$, where $\hat{f}(x) = x^3/3$. Recall also the notion of a primitive function.

9.5.2 Heat in the Carnot Cycle

Similarly, we can simply write the energy received by heat. In fact, on an isotherm with initial volume V_a, final volume V_b and condition $(V_b - V_a)\hat{\Lambda}_V(V, T) \geq 0$ the following applies

$$Q^+ = \int_{V_a}^{V_b} \Lambda_V \mathrm{d}V, \tag{9.24}$$

as implied by our heating model; on an ordinary Carnot cycle, then

$$Q^+ = \int_{\phi_1(T^+)}^{\phi_2(T^+)} \hat{\Lambda}_V(V, T^+)\,\mathrm{d}V \equiv h(T^+), \tag{9.25}$$

where we have denoted

$$\hat{h}(T) = \int_{\hat{\phi}_1(T)}^{\hat{\phi}_2(T)} \hat{\Lambda}_V(V, T)\,\mathrm{d}V. \tag{9.26}$$

9.5.3 Axioms and Carnot Cycles

What we have derived so far in this section have been the general mathematical regularities of ordinary Carnot cycles (in the rest of this section, we will not emphasize the "ordinary Carnot" as a rule). We now turn to the application of specific axioms. Axiom II has two implications. Combining it with axiom I yields $JQ > 0$, and hence $Q > 0$. It also allows us to find two cycles for which axiom III applies. Namely, it determines that the work in these cycles $(-W)$ is always a positive number, or that the integral (9.20) exists. The cycles satisfying axiom III can differ only in their adiabates, and these must be such that the integration (9.25) gives the same number for both cycles. In other words, we choose the adiabats of these cycles as follows: they intersect the isotherm T^+ at points with volume coordinates V_a and V_b, and we choose these points for different cycles so that the area under the graph of the function $\hat{\Lambda}(V, T^+)$ defined by the volumes V_a and V_b is the same for all cycles (only two are sufficient). An example is shown in Fig. 9.6.

Let us also show in the figure the choice of the two cycles from axiom III. In the top of Fig. 9.7 there are examples of two cycles with identical upper and lower temperatures, the arrow shows the progression through the cycle in time; cycle 1 has a positive, cycle 2 a negative sign of the Λ_V coefficient. However, we can choose any two "smaller" cycles lying within the illustrated cycles with common temperatures between the values of A and B, as shown in the bottom of Fig. 9.7 for temperatures T^+, T^-. Let us then call the cycle with operating temperatures A and B ("outer" cycle) the generating cycle. This lower figure also indicates the designation of the two adiabatic values in a given cycle and the four volume values that are the intersections

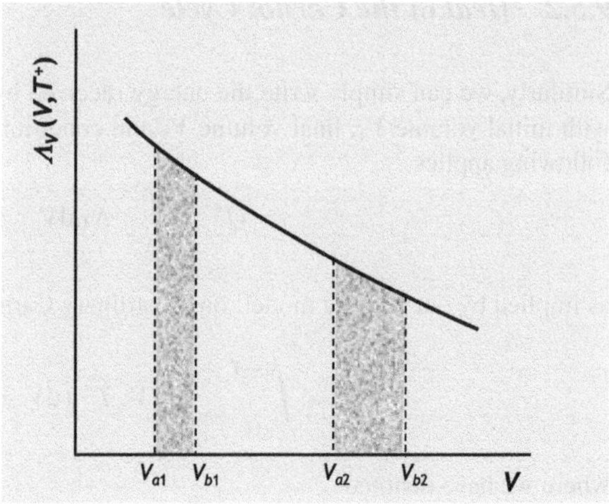

Fig. 9.6 To discussion of the implications of the axioms—the marked areas are the same, corresponding to the energy absorbed by heat at the isotherm of cycles 1 and 2. Applies to the case $\Lambda_V > 0$

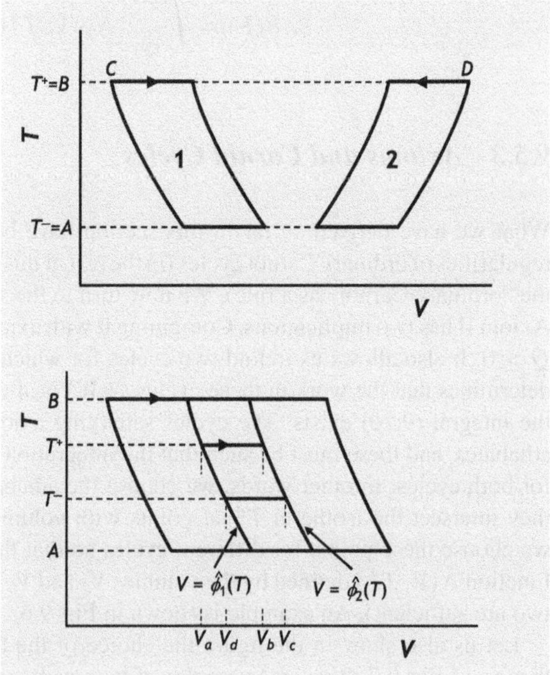

Fig. 9.7 Choice of cycles for proving the concept of entropy; see text for details

of its isotherms and adiabatic values. The generating cycle defines a closed region on which we perform mathematical operations.

In fact, axiom III implies that the efficiency of the cycles mentioned therein is the same. By efficiency we mean the ratio (proportion) of what we get (energy gained by work) to what we have to expend (energy supplied by heat). Thus, Axiom III states

$$\frac{-W_1}{Q_1^+} = \frac{-W_2}{Q_2^+} \tag{9.27}$$

(1 and 2 number the two cycles chosen, as indicated in the top of Fig. 9.7). Let us first discuss the application of the right established functions $\hat{g}$ and $\hat{h}$ to the chosen cycle, or those smaller cycles contained in it. Both functions are obviously continuously differentiable. We can express the work and the heat received with the help of (9.23) and (9.25) as differences and functions

$$-W_1 = \hat{g}_1(T^+) - \hat{g}_1(T^-), \qquad -W_2 = \hat{g}_2(T^+) - \hat{g}_2(T^-);$$

$$Q_1^+ = \hat{h}_1(T^+), \qquad Q_2^+ = \hat{h}_2(T^+). \tag{9.28}$$

Equation (9.27) of axiom III then looks like this:

$$\frac{\hat{g}_1(T^+) - \hat{g}_1(T^-)}{\hat{h}_1(T^+)} = \frac{\hat{g}_2(T^+) - \hat{g}_2(T^-)}{\hat{h}_2(T^+)}. \tag{9.29}$$

In Fig. 9.7 the lowest operating temperature of the selected cycles is marked with the letter A, the highest with the letter B. For these (common to the cycles) temperatures, Eq. (9.27) of axiom III looks like this:

$$\frac{g_1(B) - g_1(A)}{h_1(B)} = \frac{g_2(B) - g_2(A)}{h_2(B)}. \tag{9.30}$$

But we can choose any two "smaller" cycles lying within the illustrated cycles with common temperatures between the values of A and B. Thus Eq. (9.29) applies to all T^+ and T^- such that $B \geq T^+ \geq T^- \geq A$. Therefore, the following also applies

$$\frac{g_2(B) - \hat{g}_2(T^-)}{h_2(B)} = \frac{g_1(B) - \hat{g}_1(T^-)}{h_1(B)}, \tag{9.31}$$

and this for $B \geq T^- \geq A$. From Eq. (9.31) we can express the function $\hat{g}_2$:

$$\hat{g}_2(T^-) = g_2(B) - \frac{h_2(B)}{h_1(B)} g_1(B) + \frac{h_2(B)}{h_1(B)} \hat{g}_1(T^-) \equiv K\hat{g}_1(T^-) + N \tag{9.32}$$

for $A \leq T^- \leq B$ (both K and N are for the chosen B constants). Given the above implication of axiom II and the definition of Q^+ (Eq. (6.3); $\hat{h}$ is a positive function), $K > 0$. By substituting the function $\hat{g}_2$ from (9.32) to (9.29) we have

$$\frac{\hat{g}_1(T^+) - \hat{g}_1(T^-)}{\hat{h}_1(T^+)} = K \frac{\hat{g}_1(T^+) - \hat{g}_1(T^-)}{\hat{h}_2(T^+)} \tag{9.33}$$

for $B \geq T^+ \geq T^- \geq A$. Since $-W > 0$ (again, see the consequence of axiom II), by (9.28) $g_1(T^+) > g_1(T^-)$ for $T^+ \neq T^-$, so even $\hat{g}_1$ is an increasing function. Therefore, for $T^+ \neq T^-$, $g_1(T^+) = g_1(T^-)$ cannot apply, and we conclude from Eq. (9.33) that

$$\hat{h}_2(T^+) = K\hat{h}_1(T^+) \tag{9.34}$$

for $B \geq T^+ > A$. If $\hat{h}_2$ is a continuous function, (9.34) also applies for $T^+ = A$.

9.5.4 *Localization in Proof*

Finally, we are ready for the final proof under discussion. We relate the introduced functions $\hat{g}$ and $\hat{h}$ to the energy exchanged by work and heat in the cycle. We can rewrite one of the terms in (9.27) using (9.28) in the form

$$\frac{-W_1}{Q_1^+} = \frac{\hat{g}_1(T^+) - \hat{g}_1(T^-)}{\hat{h}_1(T^+)}$$

so that

$$-W_1 = \frac{\hat{g}_1(T^+) - \hat{g}_1(T^-)}{\hat{h}_1(T^+)} Q_1^+. \tag{9.35}$$

Substituting the definitions of W and Q for the ordinary Carnot cycle into (9.21) and (9.24), or (9.25) we finally get

$$\int_{T^-}^{T^+} \mathrm{d}T \int_{\hat{\phi}_{1,1}(T)}^{\hat{\phi}_{2,1}(T)} \frac{\partial \hat{P}}{\partial T} \mathrm{d}V = \frac{g_1(T^+) - g_1(T^-)}{h_1(T^+)} \int_{V_a}^{V_b} \hat{\Lambda}_V(V, T^+) \mathrm{d}V. \tag{9.36}$$

The indices next to the adiabatic curve symbol ϕ remind us that these are the first and second adiabatic curves of cycle 1 (cf. Fig. 9.7). The derivation we are looking for appears on the left-hand side in the integral. We can get to it by deriving and finding the limit, which allows us to "get rid of the integration". Thus we move from a truly integral view, looking at the whole region over which we are integrating, to a pointwise, local view. But the derivative is just a local characteristic, its value can vary from point to point. Equation (9.36) reminds us that we are dealing with a closed region bounded by a generating Carnot cycle (and we are still in the constitutive domain of some body, here quite a general one).

Let us carry out the procedure just outlined. First, let us fix the point $[V_a; T^+]$ in the domain of our cycle. Next, let us choose fixed (but arbitrary) V_b, derive (9.36) with respect to T^- and take the limit for $T^- \to T^+$ from the left. The derivative actually implies (observe Fig. 9.7) a change in the position of the "lower" isotherm of the cycle in question; the fixing of V_a and V_b determines also both adiabats, $\hat{\phi}_1(T)$

and $\hat{\phi}_2(T)$. The limit transition from T^- to T^+ from the left is $\hat{\phi}_1(T)$ to V_a, $\hat{\phi}_2(T)$ to V_b and g_1' to $g_1'(T^+)$. The result of these operations is

$$\int_{V_a}^{V_b} \frac{\partial \hat{P}(V, T^+)}{\partial T} \mathrm{d}V = \frac{g_1'(T^+)}{h_1(T^+)} \int_{V_a}^{V_b} \hat{\Lambda}_V(V, T^+) \mathrm{d}V. \tag{9.37}$$

Let us derive the last equation with respect to V_b and take the limit for $V^b \to V^a$ from the right. The result is

$$\frac{\partial \hat{P}}{\partial T} = \frac{g_1'}{h_1} \Lambda_V \tag{9.38}$$

at the point $[V_a; T^+]$. We fixed this point at the beginning, but otherwise it was chosen arbitrarily inside the closed domain defined by the generating Carnot cycle. Due to the continuity of the functions in the result (9.38), this equality also applies at the boundaries of this region. The cycles were chosen arbitrarily, their numbering does not matter, which finally gives us that

$$\frac{g'}{h} \Lambda_V = \frac{\partial \hat{P}}{\partial T}. \tag{9.39}$$

The definitions of the functions $\hat{g}$ and $\hat{h}$ imply that they are continuously differentiable on the interval $\langle A; B \rangle$. Equation (9.32) implies $\hat{g}_2'(T^-) = K\hat{g}_1'(T^-)$, and by combining with (9.34) we obtain

$$\frac{\hat{g}_2'(T^-)}{\hat{h}_2(T^+)} = \frac{K\hat{g}_1'(T^-)}{K\hat{h}_1(T^+)} = \frac{\hat{g}_1'(T^-)}{\hat{h}_1(T^+)}.$$

According to the above, this applies to all temperatures T^+, T^- in $\langle A; B \rangle$, so the following also applies

$$\frac{g_2'(B)}{h_2(B)} = \frac{g_1'(B)}{h_1(B)}. \tag{9.40}$$

Let us now choose one point on the common isotherm B, in each of our two cycles, as shown by points C and D in Fig. 9.7 above. At these points, (9.39) applies, then (9.40), so in the result the following applies

$$\left.\frac{\partial \hat{P}/\partial T}{\Lambda_V}\right|_C = \left.\frac{\partial \hat{P}/\partial T}{\Lambda_V}\right|_D. \tag{9.41}$$

The points C and D were chosen arbitrarily at temperature B and so was this temperature, so that equality (9.41) applies everywhere on some given isotherm. Thus, the fraction in it, or expression (9.1), has an invariant value at a constant temperature. And that's what we were looking for.

We can further show that the expression appearing in Eq. (9.41) is generally non-negative:

$$\frac{\partial \hat{P}/\partial T}{\Lambda_V} \geq 0. \tag{9.42}$$

We consider only ordinary points in the constitutive domain $\mathcal{D}$. Since the function $\hat{\Lambda}_V$ is continuous, Λ_V always has the same sign at all points in a sufficiently small neighbourhood of some ordinary point. This determines the adiabatic directive, see Eq. (9.8), which therefore does not change in that neighbourhood. Then each point in this neighbourhood is the intersection of one adiabatic and one isotherm, and thus we can construct a Carnot cycle in this neighbourhood. Let us assume that at that ordinary point $\Lambda_V > 0$ and $\partial \hat{P}/\partial T < 0$. Then there exist ordinary Carnot cycles around such a point, oriented clockwise[6] and containing only points where $\partial \hat{P}/\partial T < 0$. By the definition of work (9.20), the work $-W$ is then negative in these cycles, which contradicts axiom II. A similar argument can be made in the case $\Lambda_V < 0$ and $\partial \hat{P}/\partial T > 0$. We see that Λ_V and $\partial \hat{P}/\partial T$ cannot have opposite signs, as shown by the inequality in Eq. (9.42); the equation in it then applies to the case $\partial \hat{P}/\partial T = 0$.

9.6 To the Entropy via the Integration Factor

9.6.1 Motivation for the Integration Factor

We can also get to the entropy in an almost purely mathematical way. It is not that we have been discovering it non-mathematically so far, but rather we have been trying to express empirical experience mathematically. We will now show even that "more mathematical" way, because it has interesting implications for our basic study of thermal behaviour.

The right-hand side of our heating model (6.5) could be the time derivative of some function (of volume and temperature). Whether this is indeed the case we cannot tell from the model alone, but experience tells us that it will not be—then the integral of the heating with time as the integrating variable would be the integral of the total differential of some such function and would not depend on the integration path, i.e., the curve along which it is integrated. Even so, we can still turn to the theory of differential equations. This theory is also interested in the case when the expression $\hat{M}(x, y)dx + \hat{N}(x, y)dy$, called the differential form, is the total differential of some function $\hat{f}(x, y)$ [5]. In fact, if such a function exists, then $\hat{f}(x, y) = c$, where c is a constant, is a solution of the differential equation $\hat{M}(x, y)dx + \hat{N}(x, y)dy = 0$, which is obtained by simply rewriting the differential equation $dy/dx = -\hat{M}(x, y)/\hat{N}(x, y)$. Well, if such a function does not exist

---

[6] The orientation is determined by the sign of the coefficient Λ_V; for the proof we just refer the reader to the book [4].

directly, it can often be obtained by multiplying the original differential form by a suitable function $\hat{\mu}(x, y)$. We call such a function an *integration* (sometimes integrating) *factor*, because it converts the original differential form into the total differential of some function, and this is then easy to integrate (i.e., it is easy to solve the differential equation just mentioned). If the original differential form was already a total differential, the integrating factor is obviously equal to one in this case ($\mu = 1$).

9.6.2 The Integration Factor in Our Heating Model

In the theory of differential equations there is a nice and simple theorem[7]: if the functions $\hat{M}(x, y), \hat{N}(x, y)$ from the previous paragraph are continuous and if at some point $[x_0, y_0]$ applies that $M^2(x_0, y_0) + N^2(x_0, y_0) \neq 0$, then the corresponding differential form has a continuous integration factor in some neighbourhood of the point $[x_0, y_0]$ [6].

Let us now transfer this theoretical knowledge to our heating model. In this model, the roles of $\hat{M}, \hat{N}$ are played by the functions $\hat{\Lambda}_V, \hat{C}_V$, both of which are assumed to be continuous (see the text after Eq. (6.5)). Condition (6.6) is guaranteed to satisfy $M^2 + N^2 \neq 0$ actually at any point in the constitutive domain. This satisfies the conditions of the theorem stated in the previous paragraph, and our heating model thus has an integration factor in some neighbourhood of this arbitrary point. That is, if we denote this integration factor by μ, again, the expression $\mu\Lambda_V dV + \mu C_V dT$ is the total differential of some function, let us denote it by $\hat{S}(V, T)$, and then for its time derivative the following applies

$$\dot{S} = \mu Q_f = \mu(\Lambda_V \dot{V} + C_V \dot{T}). \tag{9.43}$$

So it is our old familiar entropy, albeit still with a general integration factor.

9.6.3 With an Integration Factor to Entropy

How do we find the integration factor? The relevant mathematical theory gives us a guide to this too—from the condition for a differential form to be a total differential. In our case, the condition is

$$\frac{\partial(\hat{\mu}\hat{\Lambda}_V)}{\partial T} = \frac{\partial(\hat{\mu}\hat{C}_V)}{\partial V} \tag{9.44}$$

---

[7] Originally, the theorem refers to the dependence of the solution of a differential equation on its initial conditions.

and after performing the derivations we get the equation for the integration factor:

$$\mu\left(\frac{\partial \hat{\Lambda}_V}{\partial T} - \frac{\partial \hat{C}_V}{\partial V}\right) = C_V \frac{\partial \hat{\mu}}{\partial V} - \Lambda_V \frac{\partial \hat{\mu}}{\partial T}. \tag{9.45}$$

Let us first consider the case where the integration factor is just a function of temperature. This would imply fulfilling the condition, arising from Eq. (9.45), that the fraction

$$\frac{\dfrac{\partial \hat{\Lambda}_V}{\partial T} - \dfrac{\partial \hat{C}_V}{\partial V}}{-\Lambda_V} \tag{9.46}$$

is also just a function of temperature. The integration factor would then be determined from the modified Eq. (9.45):

$$\frac{d \ln \mu}{dT} = \frac{\dfrac{\partial \hat{\Lambda}_V}{\partial T} - \dfrac{\partial \hat{C}_V}{\partial V}}{-\Lambda_V} \tag{9.47}$$

(so we only consider ordinary points). By substituting from the equations of state and functions of the ideal gas and of Van der Waals fluid into expression (9.46) we find that the fraction has a common value, $-1/T$, which is just a function of temperature. We can argue that we are looking for a universal integration factor for all fluids, which is thus a function of temperature only. Or we note that the numerator of the fraction (9.46) is equal to $(1/J)\partial \hat{P}/\partial T$ according to Eq. (8.7), and then the whole fraction is equal to $-1/T$ according to the universally valid Eq. (9.3). Thus, if we substitute the expression $-1/T$ into the right-hand side of Eq. (9.47) we determine the value of the integration factor $1/T$. The first Eq. (9.43) then reads

$$\dot{S} = Q_f/T, \tag{9.48}$$

which corresponds to the previous derivation of entropy. This relationship is called the *second law of thermodynamics* in the form of (entropic) equality. Thus, the existence of entropy is a consequence of the existence of an integrating heating factor. The entropy itself is not the integrating factor, that one is the inverse temperature $1/T$. We have derived the entropy in this way only locally, i.e., in a certain neighbourhood of the (ordinary) point $[x_0, y_0]$ or $[V_0, T_0]$, but the derivation shows that this is sufficient.

9.7 Integration Factor and Efficiency

9.7.1 *Efficiency of Heat Engines*

The existence of the heating integration factor has interesting implications for the assessment of the efficiency of energy transfer in thermal machines. They take in energy through heat and we can get energy back from them through work. Efficiency is then determined by the ratio of "return" to "cost". Heat, the energy supplied by the heat, is what we have to expend to operate the machine: Q^+. What we gain is the net work during the work cycle $(-W)$, but what we may lose is the energy emitted by the heat: Q^-. The basis for assessing the efficiency is therefore the comparison of the energy absorbed and emitted by the heat in the cycle.

This is illustrated by the first law of thermodynamics, the first Eq. (7.14). If the internal energy were constant, then after integration from the initial to the final time of the relevant process, we get $Q = -W$, i.e., that all the energy exchanged by heat is equivalent to work. The internal energy is constant, for example, right in the cycle, more precisely, the integral of $\dot{U}$ is zero in the cycle. But the heat is given by the difference of two positive numbers:

$$Q = Q^+ - Q^- \tag{9.49}$$

as implied by definition (6.1) and equations (6.3), (6.4). If the heat emitted is not zero, it reduces the value of Q, for a given value of heat absorbed, and thus also the value of work recoverable under the conditions just discussed. Paying attention not only to the total integral of heating in a given process (cycle), but to its "division" into absorbed and released heat, is therefore very important. This was pointed out at the end of Sect. 9.4.

9.7.2 *Heating and its Integration Factor in a Cycle*

If the reciprocal (absolute) temperature is the integration factor of the heating (in our heating model), then in a cycle that starts at time t_1 and ends at time t_2, the following applies:

$$\oint_{t_1}^{t_2} \frac{Q_f}{T} dt = \oint_{t_1}^{t_2} \dot{S} dt = 0 \tag{9.50}$$

The heating integral appeared in Eq. (6.1) and the following. Here the heating is further multiplied by the inverse of the temperature. It is not a problem to introduce such a multiplier into the relations with heating. Consider some integrable function f and use the division of the time interval into $\mathcal{T}^+$ and $\mathcal{T}^-$, see (9.19); then we can write

$$\int_{\mathcal{T}^+} f Q_f \mathrm{d}t \equiv \frac{1}{2}\int_{t_1}^{t_2} (|Q_f| + Q_f) f \mathrm{d}t, \tag{9.51}$$

$$\int_{\mathcal{T}^-} f Q_f \mathrm{d}t \equiv -\frac{1}{2}\int_{t_1}^{t_2} (|Q_f| - Q_f) f \mathrm{d}t \tag{9.52}$$

or for energy absorbed or emitted by heat:

$$Q^+ = \int_{\mathcal{T}^+} Q_f\, \mathrm{d}t, \qquad Q^- = -\int_{\mathcal{T}^-} Q_f\, \mathrm{d}t. \tag{9.53}$$

Let's put $f = 1/T$ into Eqs. (9.51) and (9.52), combine them with Eq. (9.50); then we have the following equations in the cycle:

$$\int_{\mathcal{T}^+} \frac{1}{T} Q_f \mathrm{d}t = \frac{1}{2}\int_{t_1}^{t_2} \frac{1}{T}|Q_f| \mathrm{d}t,$$

$$\int_{\mathcal{T}^-} \frac{1}{T} Q_f \mathrm{d}t = -\frac{1}{2}\int_{t_1}^{t_2} \frac{1}{T}|Q_f| \mathrm{d}t$$

and by adding them together we get

$$\int_{\mathcal{T}^+} \frac{1}{T} Q_f \mathrm{d}t + \int_{\mathcal{T}^-} \frac{1}{T} Q_f \mathrm{d}t = 0. \tag{9.54}$$

Heating is associated with the heat absorbed and given off according to Eq. (9.53). These can still be multiplied by arbitrary constants, one by constant p, the other by q, subtracted and zero added to them in the form (9.54)

$$\begin{aligned} qQ^- - pQ^+ &= -\int_{\mathcal{T}^-} qQ_f\, \mathrm{d}t - \int_{\mathcal{T}^+} pQ_f\, \mathrm{d}t + \int_{\mathcal{T}^+} \frac{1}{T} Q_f \mathrm{d}t + \int_{\mathcal{T}^-} \frac{1}{T} Q_f \mathrm{d}t \\ &= \int_{\mathcal{T}^+} Q_f \Big(\frac{1}{T} - p\Big) \mathrm{d}t + \int_{\mathcal{T}^-} (-Q_f) \Big(q - \frac{1}{T}\Big) \mathrm{d}t \end{aligned} \tag{9.55}$$

Well, Eq. (9.55) allows us to consider the two types of heat in our cycle. Our integration factor, $1/T$, contains temperature as a continuous function of time. This function has a maximum and a minimum on the interval $\langle t_1; t_2\rangle$ which we simplistically denote by $T_{\max}$, and $T_{\min}$, respectively. If $\mathcal{T}^+$ is non-empty, this function has a finite supremum and infimum on it, we denote them by $\sup T_{\mathcal{T}^+}$ and $\inf T_{\mathcal{T}^+}$. This is analogous for the non-empty interval $\mathcal{T}^-$, where we denote by $\sup T_{\mathcal{T}^-}$ and $\inf T_{\mathcal{T}^-}$.

Let us now choose the constants p and q in Eq. (9.55) as follows: $p = 1/T_{\max}$ and $q = 1/T_{\min}$. Then the integrands in Eq. (9.55) are nonnegative. Thus $Q^- \geq (T_{\min}/T_{\max})Q^+$, and the equality applies if and only if both integrands are zero, that is, if and only if $T = T_{\max}$ on $\mathcal{T}^+$ and $T = T_{\min}$ on $\mathcal{T}^-$. If we swap the choice of p and q, we get similarly: $Q^- \leq (T_{\max}/T_{\min})Q^+$. Thus, in summary, we get the following for the comparison of the two heats in the cycle:

$$(T_{\max}/T_{\min})Q^+ \geq Q^- \geq (T_{\min}/T_{\max})Q^+ \tag{9.56}$$

with the above caveat for the validity of the equal signs.

Similarly, we can argue with supremes and infimas, introduced two paragraphs above; we arrive at inequalities:

$$(\sup T_{\mathcal{T}^-}/\inf T_{\mathcal{T}^+})Q^+ \geq Q^- \geq (\inf T_{\mathcal{T}^-}/\sup T_{\mathcal{T}^+})Q^+, \tag{9.57}$$

where the equal signs apply if and only if T is constant on both $\mathcal{T}^+$ and $\mathcal{T}^-$.

9.7.3 *Efficiency is Limited*

From Eq. (9.56) it follows that in a cyclic process $Q^- = 0$ if and only if $Q^+ = 0$. In other words: $\mathcal{T}^+$ is empty if and only if $\mathcal{T}^-$ is empty. In practical terms, this means that we cannot avoid losses (non-zero Q^-), hence we cannot achieve 100% efficiency. However, we have excluded the (unattainable) zero (absolute) temperature from these considerations; at that temperature the losses could be zero, as indicated by both Eqs. (9.56) and (9.57). Yet another way to formulate our reasoning about the impact of these equations is as follows: in a cyclic process, a body cannot absorb heat without also emitting heat. This statement serves in the more advanced and general theory as a different formulation of the second theorem and as the starting point of the proof of the existence of entropy [7, 8]; in the simpler theory interpreted here, it follows as a consequence of the existence of the integration factor. We can see that even in our basic, simple model of heating, where we see no "loss" of energy to the body transferred by heat, there is still a loss, in the sense of the existence of emitted heat that reduces the work done from the absorbed heat (in the cycle). Thus, the existence of the integration factor has two consequences—the existence of a function called entropy and the proof of the impossibility to achieve in the cyclic process the conversion of all the energy obtained by heat into energy given up by work, because always a part of the energy is given up by heat. The latter in practice we understand as (thermal) energy loss.

9.8 The Meaning of Entropy

Entropy is primarily—unlike heating or heat—a function, in our basic form, of temperature and volume. It shows how to convert heating into just such a function; this is handled by the integration factor. A function representing something that is conserved (unchanged) in the cycle apart from (internal) energy and that is closely related to heat. Heat is generally not conserved in the cycle; the integral of heating in

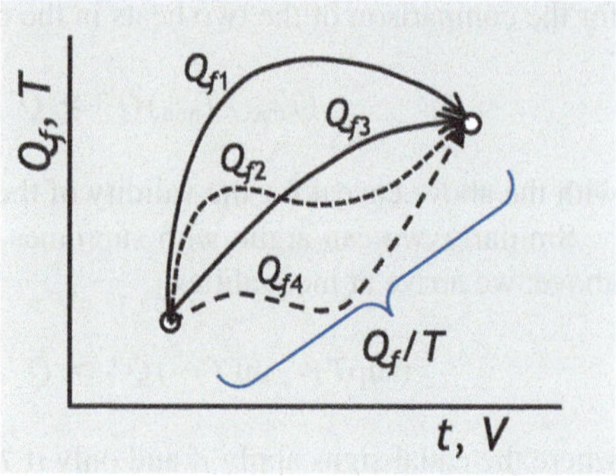

Fig. 9.8 The value of heat depends on the course of process, the integral of entropy does not

the cycle is not zero. However, the integral of the heating multiplied by the reciprocal (absolute) temperature in the cycle is zero and Q_f/T thus represents the quantity that is conserved in the cycle. This quantity is equal to the time derivative of entropy in our basic heating model, its integral in the cycle is zero, so entropy is conserved in the cycle, making it related to energy. Thus a cyclically operating heat machine does not "consume" the Q_f/T quantity, even though it needs a constant supply of heat energy to operate. Similarly, a water mill does not consume the water that drives its wheel, it just uses its kinetic energy. It is interesting that two quantities describing thermal behaviour (heat, or heating, and temperature) are thus combined in entropy. Entropy appears here as a consequence of the existence of the integrating factor of heating.

This is illustrated schematically in Fig. 9.8. The energy exchanged by heat, given by the heating integral, may be different for different paths (processes) between the same points, but the integral of the heating and temperature fraction is still the same for all paths.

The entropy is also the function determining in our heating model its two constitutive coefficients—see Eq. (8.15). We can understand this to mean that the entropy as a function determines how the energy transferred by heat in a given body manifests itself in a change in its temperature and/or volume; it determines the heat constitutive equations. Although the heating is not a function (of temperature and volume) and its integral depends on the curve of action connecting two given points (given initial and final points of integration), these coefficients are such a function and, moreover, both are determined by partial derivatives of another, thus somehow "superior" function—entropy.

And yet in other words. Our model (6.5) does not specify heating as a function of volume and temperature, although two such functions appear in it. It also includes functions of time, or their derivatives. By integrating the heating, we do get heat, the heating is an (intrinsic) function of time, but this heat will depend not only on the volume and temperature, but also on their time course, i.e., on the curve in the time coordinates along which it is integrated. Entropy is the result of converting heating to a function of volume and temperature. Simplistically, we can say that entropy is the conversion of heating from a "mere" function of time to a function of volume and temperature, directly related to the body. It is a kind of constitutive transducer that determines the two constitutive parameters in the heating model.

Let us illustrate these considerations mathematically. The process $\mathcal{P}$ is represented by a curve, i.e., a set of points $V(t), T(t), t_1 \le t \le t_2$. It is known from calculus that for the line integral of the function $\hat{S}(V, T)$ along this curve, the following applies

$$\int_{\mathcal{P}} \left[\frac{\hat{\Lambda}_V(V,T)}{T}\,dV + \frac{\hat{C}_V(V,T)}{T}\,dT\right] = \int_{t_1}^{t_2} \left[\frac{\hat{\Lambda}_V(\hat{V}(t),\hat{T}(t))}{T}\dot{V} + \frac{\hat{C}_V(\hat{V}(t),\hat{T}(t))}{T}\dot{T}\right]dt. \quad (9.58)$$

However, on the left-hand side we have the integral of $(Q_f/T)\,dt$ and on the right-hand side the integral of the total differential entropy as a function of time, which is equal to $\Delta S = S(t_2) - S(t_1)$.

In the case of constant temperature, where according to (6.5) all the energy supplied by heat goes to volume change, i.e., potentially "to work" (volume), we have

$$\Delta S_{\text{izot}} = \int_{\mathcal{P}} \frac{\hat{\Lambda}_V}{T}\,dV = \int_{t_1}^{t_2} \frac{Q_f}{T}\,dt = \frac{Q_{\text{izot}}}{T} \quad (9.59)$$

and the entropy change multiplied by the corresponding temperature is directly equal to the energy exchanged by heat; the heat then does not depend on the path of the isothermal process. Furthermore, it is clear that if a body absorbs energy by heat, the entropy increases; if it emits, the entropy decreases. The connection with work is no longer so straightforward. From the heating model (6.5) and the relation $Q_f = T\dot{S}$ for the case T = const. it implies (at the ordinary point) $\dot{V} = T\dot{S}/\Lambda_V$ and, when substituted into the definition of the work (7.2):

$$-W_{\text{izot}} = T\int_{t_1}^{t_2} \frac{P}{\Lambda_V}\dot{S}\,dt. \quad (9.60)$$

The integral on the right-hand side cannot be expressed or quantified, in general. In the case of an ideal gas, where $\Lambda_V = P$ then we get a direct and simple relationship between the energy exchanged by work and the change in entropy (again multiplied by temperature). At the same time, it implies a complete equivalence between heat and work, $Q_{\text{izot}} = -W_{\text{izot}}$, and this in a non-cyclic process.

In the isochoric case, when all the heat goes to the temperature change (i.e., work is zero), we get

$$\Delta S_{\text{izoch}} = \int_{\mathcal{P}} \frac{\hat{C}_V}{T}\,dT \quad (9.61)$$

and without knowing the temperature dependence of C_V we cannot integrate the right-hand side further; however, it is clear that the integral between given points will not depend on the curve $\mathcal{P}$, that connects them (the process that connects them), the integral is given by the corresponding entropy change. But if the heat capacity does not depend on the temperature, at least within our process, i.e., in the interval $\langle t_1; t_2\rangle$, it is valid

$$\Delta S_{\text{izoch}} = C_V \int_{t_1}^{t_2} \frac{dT}{T} = C_V \ln \frac{T(t_2)}{T(t_1)} \quad (C_V = \text{const.}). \tag{9.62}$$

We have figured that here the heating goes just to a temperature change. Since (6.6), applies, then the last equation says that entropy also increases as temperature increases and vice versa. We can follow a very similar procedure for the constant pressure process using (6.11). The procedure is almost identical, with the resulting equations very close to the last two, so we leave this short derivation to our readers.

We can still rewrite the integral on the right-hand side of (9.58):

$$\int_{t_1}^{t_2} \frac{\Lambda_V}{T} \dot{V} dt = \Delta S - \int_{t_1}^{t_2} \frac{C_V}{T} \dot{T} dt. \tag{9.63}$$

If we consider the left-hand side to be the part of the heating "used for work", then the right-hand side is a kind of measure of the work obtainable from the heating, and this measure is affected by the change in entropy (not just the temporal change in temperature). So, this is another demonstration of the place of entropy as a function describing energy transformations by heat and work.

What entropy as a function (of temperature and volume) specifically looks like is not seen here. Yet, as we have just illustrated, we can calculate its change in various cases. This is often sufficient, after all we are in a similar position for internal energy.

Beware of confusing entropy with the heat gained in a given process [9]—this is the integral of heating from initial to final time, which is energy, while entropy is energy per unit temperature. The relationship between the two quantities is shown by integrating Eq. (9.48) per parties:

$$Q = \int_{t_0}^{t} Q_f dt = [TS]_{t_0}^{t} - \int_{t_0}^{t} \dot{T} S dt \tag{9.64}$$

References

1. Malcolm, G.N., Ritchie, G.L.D.: Direct measurement of the thermal pressure coefficient of water. Aust. J. Chem. **15**(4), 740–743 (1962)
2. Linstrom, P.J., Mallard, W.G. (eds.): NIST Chemistry WebBook, NIST Standard Reference Database Number 69. National Institute of Standards and Technology, Gaithersburg. https://doi.org/10.18434/T4D303 (visited June 6, 2022)
3. Carnot, S.: Reflections on the Motive Power of Fire. Dover, Mineola (1988) (reprint of English translation)
4. Truesdell, C., Bharatha, S.: The Concepts and Logic of Classical Thermodynamics as a Theory of Heat Engines, Rigorously Constructed upon the Foundation Laid by S. Carnot and F. Reech. Springer, New York (1977)

5. Katz, V.J.: The history of differential forms from Clairaut to Poincaré. Hist. Math. **8**(2), 161–188 (1981)
6. Petrovski, I.G.: Ordinary Differential Equations. Prentice-Hall, Englewood Cliffs (1966)
7. Pekař, M., Samohýl, I.: The Thermodynamics of Linear Fluids and Fluid Mixtures. Springer, Cham (2014). See also https://www.fch.vut.cz/vav/sapelite
8. Šilhavý, M.: The Mechanics and Thermodynamics of Continuous Media. Springer, Berlin, Heidelberg (1997)
9. Truesdell, C.: Rational Thermodynamics, 2nd edn. Springer, New York (1984)

Chapter 10
Thermodynamic Potentials. Essentials Overview

Abstract The concept of thermodynamic potentials is explained. The most important potentials are then derived carefully operating on functions expressing thermodynamic quantities as functions of various pairs of independent variables. The demonstrated potentials are Helmholtz energy, Gibbs energy, internal energy, and enthalpy. Reciprocal relations and their practical significance are mentioned, as well as links between Helmholtz energy and work or between enthalpy or internal energy and heat. The whole structure of classical (equilibrium) thermodynamics can be constructed on the basis of several fundamental principles—axioms rooted in experience with thermal behaviour. The principles are summarized in the last section.

In this chapter we attempt to give a mathematically correct procedure for what is usually done rather sloppily in traditional thermodynamics—substituting independent variables and expressing partial derivatives of related functions. The derivations will be valid in a domain where the relevant mathematical operations, in particular function inversions, can be performed and which are appropriately mathematically coherent in terms of the existence of the functions found. We will not discuss such a domain in detail, contenting ourselves with the statement that it is a sufficiently large part of the constitutive domain for practical purposes.

10.1 Meaning of Thermodynamic Potentials; Helmholtz Energy

From the second Eq. (7.14), we can express the pressure as a function as follows (we are already working with a unit value for J):

$$\hat{P} = \Lambda_V - \frac{\partial \hat{U}}{\partial V} = T\frac{\partial \hat{S}}{\partial V} - \frac{\partial \hat{U}}{\partial V}. \tag{10.1}$$

On the right-hand side of Eq. (10.1) we can see the partial derivatives of two different volume (and temperature) functions. It would be simpler to have the derivative of a

© The Author(s), under exclusive license to Springer Nature Switzerland AG 2024

M. Pekař, *The Essentials of Thermodynamics*, Physical Chemistry in Action,
https://doi.org/10.1007/978-3-031-60321-1_10

single function there to determine the pressure thus occurring on the left side of this equation. This is easily achieved if we define the new function by the relation[1]:

$$\hat{F}(V,T) = \hat{U}(V,T) - T\hat{S}(V,T). \tag{10.2}$$

Thus its partial derivative with respect to volume looks like this:

$$\frac{\partial \hat{F}}{\partial V} = \frac{\partial \hat{U}}{\partial V} - T\frac{\partial \hat{S}}{\partial V}. \tag{10.3}$$

Combining Eqs. (10.1) and (10.3) we get what we were looking for:

$$\hat{P} = -\frac{\partial \hat{F}}{\partial V}. \tag{10.4}$$

Let's also explore the higher derivatives of the new function—the astute reader will note that we can use them to relate it to two other characteristics of the body—the coefficients of our thermal model, see (8.15). The second mixed derivative looks like this:

$$\frac{\partial^2 \hat{F}}{\partial T \partial V} = \frac{\partial^2 \hat{U}}{\partial T \partial V} - \frac{\partial \hat{S}}{\partial V} - T\frac{\partial^2 \hat{S}}{\partial T \partial V}. \tag{10.5}$$

After substitution from the second equality (7.14) and from (8.15) we get

$$\frac{\partial^2 \hat{F}}{\partial T \partial V} = \frac{\partial \hat{C}_V}{\partial V} - \frac{\Lambda_V}{T} - \frac{\partial \hat{C}_V}{\partial V}, \tag{10.6}$$

from which we can finally deduce

$$\Lambda_V = -T\frac{\partial^2 \hat{F}}{\partial T \partial V} \tag{10.7}$$

(the order of the partial derivative, as we know, can be reversed). Thus, the new function (its derivative) also determines Λ_V.

And now the partial derivative with respect to temperature:

$$\frac{\partial \hat{F}}{\partial T} = \frac{\partial \hat{U}}{\partial T} - \hat{S} - T\frac{\partial \hat{S}}{\partial T}, \tag{10.8}$$

using the second equality (8.15):

$$\frac{\partial^2 \hat{F}}{\partial T^2} = \frac{\partial^2 \hat{U}}{\partial T^2} - \frac{\partial \hat{S}}{\partial T} - \frac{\partial \hat{C}_V}{\partial T}. \tag{10.9}$$

[1] A definition with opposite signs would probably be more natural for our procedure, but the history of thermodynamics has taken that definition.

From the second equality (7.14) we can see that the first and third terms of the right-hand side of the last equation cancel out, and we can add the second term from the second equality (8.15) to obtain finally:

$$C_V = -T\frac{\partial^2 \hat{F}}{\partial T^2}. \tag{10.10}$$

The new function (its) derivative finally determines C_V as well.

Let us repeat what we have obtained in the last paragraphs. We have discovered a function of volume and temperature which alone determines the characteristic quantities (functions) of the body in our model of thermal behaviour—pressure, heat capacity and coefficient of thermal change of volume, i.e., the constitutive quantities for which we need to know the constitutive equation. More precisely, in our case, we determine them through their derivatives. Such functions are important for this reason, they are often called (thermodynamic) potentials. The function defined by Eq. (10.2) bears various names—*Helmholtz energy*, Helmholtz potential, *free energy*. Because of the way it was derived, we obtained it—as entropy—for a simply continuous, open subset of the constitutive domain consisting only of thermodynamic points.

For the Helmholtz energy, we will show one more bonus. Substituting into Eq. (10.8) from Eqs. (8.15) and (7.14), we finally get that the Helmholtz energy also determines the entropy:

$$\hat{S} = -\frac{\partial \hat{F}}{\partial T}. \tag{10.11}$$

10.2 Gibbs Energy

There are various thermodynamic potentials. Let us note a few more that are among the most important. Helmholtz energy functioned as a potential when the thermodynamic quantities were functions of our most convenient independent variables, volume and temperature. In practice, however, instead of volume as the independent variable, pressure is often more popular. We have already said that this is a non-universal and problematic choice, however, let us consider the case where we replace the volume-temperature variables with pressure-temperature variables. So effectively we only substitute one independent variable, temperature remains in that role. How do we do this? We transform the function (6.7) (invert for volume) into the function $V = \check{V}(P, T)$. Then we can convert some function of volume and temperature $\hat{f}(V, T)$ into new variables:

$$\hat{f}(V, T) \equiv \hat{f}(\check{V}(P, T), T) \equiv \check{f}(P, T). \tag{10.12}$$

When can we make this transformation? If the conditions for the existence of an inverse function are satisfied, which is true where (6.8) applies. The relations between the partial derivatives of the functions in (10.12) follow from the chain rule:

$$\frac{\partial \hat{f}}{\partial V}\frac{\partial \check{V}}{\partial P} = \frac{\partial \check{f}}{\partial P}; \quad \frac{\partial \hat{f}}{\partial V}\frac{\partial \check{V}}{\partial T} + \frac{\partial \hat{f}}{\partial T} = \frac{\partial \check{f}}{\partial T}. \tag{10.13}$$

We did not change the temperature as an independent variable, we will probably not find an incentive for the formulation of a new potential, however, for the conversion of the derivatives according to the volume-pressure there is a factor $\partial \check{V}/\partial P$, which could inspire us to insert the product PV into the original definition of the Helmholtz energy (who does not feel this inspiration, will understand around the next equation); let's try it and denote the result by G:

$$G = U - TS + PV. \tag{10.14}$$

Let us first express the partial derivative of G as a function of the new variables with respect to temperature:

$$\begin{aligned}\frac{\partial \check{G}}{\partial T} = \frac{\partial \check{U}}{\partial T} - S - T\frac{\partial \check{S}}{\partial T} + P\frac{\partial \check{V}}{\partial T} = \frac{\partial \hat{U}}{\partial V}\frac{\partial \check{V}}{\partial T} + \frac{\partial \hat{U}}{\partial T} - S -\\ T\frac{\partial \hat{S}}{\partial V}\frac{\partial \check{V}}{\partial T} - T\frac{\partial \hat{S}}{\partial T} + P\frac{\partial \check{V}}{\partial T} = (\Lambda_V - P)\frac{\partial \check{V}}{\partial T} + C_V - S -\\ \Lambda_V\frac{\partial \check{V}}{\partial T} - C_V + P\frac{\partial \check{V}}{\partial T} = -S.\end{aligned} \tag{10.15}$$

(we used Eqs. (7.14) and (8.15)). Now, let us rewrite what we just found out:

$$S = -\frac{\partial \check{G}}{\partial T}. \tag{10.16}$$

The function $\check{G}$ therefore determines the entropy. This is very similar to Eq. (10.11), which we expected, since we have left temperature as the independent variable here. Looking beyond the second term in Eq. (10.15), we can see that in order to end up with entropy only, it was crucial to add just the product PV to the $U - TS$ difference; this was the inspiration mentioned above.

Very similarly, we could proceed with the derivative according to the second independent variable, i.e., with $\partial \check{G}/\partial P$. Leaving this as an exercise for the readers, they should arrive at the result:

$$V = \frac{\partial \check{G}}{\partial P}, \tag{10.17}$$

which is analogous to relation (10.4). The function $\check{G}$ thus determines the volume, similar to the way the function $\hat{F}$ determined the pressure.

But we still need to find out if the function $\check{G}$ is a thermodynamic potential, i.e., if it (along with temperature) determines the quantities Λ_V and C_V. The former is given by the first Eq. (8.15), and now we would need to convert the partial derivative occurring in it into a derivative of the pressure and temperature functions. We can see how this is done in Eq. (10.13), with the help of which we can write

$$\Lambda_V = T\frac{\partial \hat{S}}{\partial V} = -T\frac{\dfrac{\partial \check{S}}{\partial P}}{\dfrac{\partial \check{V}}{\partial P}} = -T\frac{\dfrac{\partial^2 \check{G}}{\partial P \partial T}}{\dfrac{\partial^2 \check{G}}{\partial P^2}}. \tag{10.18}$$

The coefficient of thermal volume change is thus effectively determined by the function $\check{G}$ (and the temperature).

From similar sources we derive

$$\begin{aligned} C_V = T\frac{\partial \hat{S}}{\partial T} &= T\left(\frac{\partial \check{S}}{\partial T} - \frac{\partial \hat{S}}{\partial V}\frac{\partial \check{V}}{\partial T}\right) = T\left(-\frac{\partial^2 \check{G}}{\partial T^2} - \frac{\Lambda_V}{T}\frac{\partial^2 \check{G}}{\partial T \partial P}\right) \\ &= -T\frac{\partial^2 \check{G}}{\partial T^2} - \Lambda_V\frac{\partial^2 \check{G}}{\partial T \partial P}. \end{aligned} \tag{10.19}$$

Thus, even to determine the heat capacity at constant volume, knowledge of the $\check{G}$ function and the temperature is sufficient (the coefficient of thermal change of volume has already been determined in the previous equation). This function is therefore the thermodynamic potential; it is called *Gibbs energy* or *free enthalpy*.[2]

Let us return to Eq. (10.18). In deriving it, we assumed the non-zero derivative we were dividing. We will now prove this fact. We start with the differential pressure as a function of the "initial" variables (volume, temperature), to which we add the inverse of the volume function:

$$\begin{aligned} \mathrm{d}P &= \frac{\partial \hat{P}}{\partial V}\mathrm{d}V + \frac{\partial \hat{P}}{\partial T}\mathrm{d}T = \frac{\partial \hat{P}}{\partial V}\left(\frac{\partial \check{V}}{\partial T}\mathrm{d}T + \frac{\partial \check{V}}{\partial P}\mathrm{d}P\right) + \frac{\partial \hat{P}}{\partial T}\mathrm{d}T \\ &= \frac{\partial \hat{P}}{\partial V}\frac{\partial \check{V}}{\partial T}\mathrm{d}T + \frac{\partial \hat{P}}{\partial V}\frac{\partial \check{V}}{\partial P}\mathrm{d}P + \frac{\partial \hat{P}}{\partial T}\mathrm{d}T \\ &= \frac{\partial \hat{P}}{\partial V}\frac{\partial \check{V}}{\partial P}\mathrm{d}P + \left(\frac{\partial \hat{P}}{\partial V}\frac{\partial \check{V}}{\partial T} + \frac{\partial \hat{P}}{\partial T}\right)\mathrm{d}T. \end{aligned} \tag{10.20}$$

Comparing the left and right sides, we can see that

$$\frac{\partial \hat{P}}{\partial V}\frac{\partial \check{V}}{\partial P} = 1 \tag{10.21}$$

[2] According to Partington [1], from the Greek $\theta\alpha\lambda\pi o\varsigma$ meaning heat; the naming will become clearer when we encounter enthalpy as such, without attributes (Sect. 10.4).

and

$$\frac{\partial \hat{P}}{\partial V}\frac{\partial \check{V}}{\partial T}+\frac{\partial \hat{P}}{\partial T}=0. \tag{10.22}$$

Considering Eq. (6.8), the first equation gives us the sought-after non-zero derivative.

The mindful reader will have raised the question of how we invert here without discretion to pressure as the independent variable, since we have already said that this is a problematic way to go. Well, here too we must proceed with caution and not immediately generalize everything. Of course, the inversion condition (6.8) applies (only) locally, i.e., in a sufficiently small neighbourhood of the point at which the derivative is evaluated. As a partial derivative, it is performed at a constant value of the second variable, the temperature. The negative sign informs that the pressure here is a decreasing function of the volume. From mathematics we know that a function on a decreasing (increasing) interval is clearly invertible. Thus, a sufficient condition for volume-pressure inversion not only locally, but everywhere in a given domain, would be that with every two distinct points of that domain lying on some isotherm of it, the entire isothermal line of those two points (the entire interval of the isotherm of which those two points are the extremes) would lie within it. Truesdell and Bharatha call such a domain isothermally convex [2].

The Van der Waals fluid is an example where this is not quite the case. In Fig. 6.4 we can see that the isotherms intersect the isobars lying below the $P_r = 1$ at two points the isotherm line of which does not lie in the constitutive domain. On such an isotherm, although condition (6.8) is satisfied, there are two functions $\check{V}(P, T)$—one for the liquid, the other for the gaseous domain. The inversion here is not unambiguous. Figure 10.1 shows data for the carbon dioxide isotherm at 273,15 K (from the NIST database [3]). It shows well the two branches, this time in pressure-volume coordinates, separated by the problematic domain.

10.3 Internal Energy as Potential

Equations (7.14) and (8.15) show the close relationship between internal energy and entropy as functions. For example, they show that

$$\frac{\partial \hat{U}}{\partial T}=T\frac{\partial \hat{S}}{\partial T} \tag{10.23}$$

and this in turn implies the relation

$$T=\frac{\dfrac{\partial \hat{U}}{\partial T}}{\dfrac{\partial \hat{S}}{\partial T}},$$

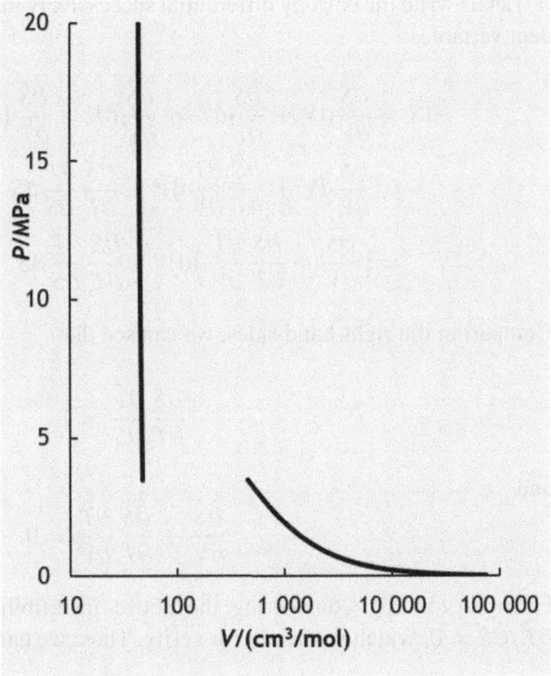

Fig. 10.1 Isotherm 273.15 K for liquid (left) and gaseous (right) carbon dioxide

which suggests that we could express the internal energy as a function of entropy. The second relation of Eq. (8.15), which implies $\partial\hat{S}/\partial T > 0$ confirms that our reasoning is sound. We can therefore perform the inversion of $S = \hat{S}(V, T)$ for temperature: $T = \bar{T}(V, S)$. Specifically for internal energy, we then change the independent variables as follows:

$$U = \hat{U}(V, T) \equiv \hat{U}(V, \bar{T}(V, S)) \equiv \bar{U}(V, S). \tag{10.24}$$

Equation (10.24) implies

$$\frac{\partial\bar{U}}{\partial S} = \frac{\partial\hat{U}}{\partial T}\frac{\partial\bar{T}}{\partial S}$$

and from this:

$$\frac{\partial\hat{U}}{\partial T} = \frac{\dfrac{\partial\bar{U}}{\partial S}}{\dfrac{\partial\bar{T}}{\partial S}},$$

unless the partial derivative in the denominator of the last fraction is zero. We will verify this again by the familiar procedure of converting partial derivatives with respect to different variables.

Let us write the entropy differential successively in the original and new independent variables:

$$
\begin{aligned}
\mathrm{d}S &= \frac{\partial \hat{S}}{\partial V}\mathrm{d}V + \frac{\partial \hat{S}}{\partial T}\mathrm{d}T = \frac{\partial \hat{S}}{\partial V}\mathrm{d}V + \frac{\partial \hat{S}}{\partial T}\Big(\frac{\partial \bar{T}}{\partial V}\mathrm{d}V + \frac{\partial \hat{T}}{\partial S}\mathrm{d}S\Big) \\
&= \frac{\partial \hat{S}}{\partial V}\mathrm{d}V + \frac{\partial \hat{S}}{\partial T}\frac{\partial \bar{T}}{\partial V}\mathrm{d}V + \frac{\partial \hat{S}}{\partial T}\frac{\partial \bar{T}}{\partial S}\mathrm{d}S \\
&= \Big(\frac{\partial \hat{S}}{\partial V} + \frac{\partial \hat{S}}{\partial T}\frac{\partial \bar{T}}{\partial V}\Big)\mathrm{d}V + \frac{\partial \hat{S}}{\partial T}\frac{\partial \bar{T}}{\partial S}\mathrm{d}S.
\end{aligned}
\tag{10.25}
$$

Comparing the right-hand sides, we can see that

$$
\frac{\partial \hat{S}}{\partial T}\frac{\partial \bar{T}}{\partial S} = 1 \tag{10.26}
$$

and

$$
\frac{\partial \hat{S}}{\partial V} + \frac{\partial \hat{S}}{\partial T}\frac{\partial \bar{T}}{\partial V} = 0. \tag{10.27}
$$

From Eq. (10.26), considering the above invertibility condition, it follows that $\partial \bar{T}/\partial S \neq 0$, which we needed to verify. Thus, we can finally write

$$
C_V = \frac{\partial \hat{U}}{\partial T} = \frac{\dfrac{\partial \bar{U}}{\partial S}}{\dfrac{\partial \bar{T}}{\partial S}} \equiv \bar{C}_V. \tag{10.28}
$$

The internal energy thus determines (as a function volume and entropy) the heat capacity at constant volume. In the last equation we can substitute from the second Eq. (8.15) and further modify using (10.26):

$$
C_V = T\frac{\partial \hat{S}}{\partial T} = \frac{\dfrac{\partial \bar{U}}{\partial S}}{\dfrac{\partial \bar{T}}{\partial S}} \Rightarrow T = \frac{\partial \bar{U}}{\partial S}. \tag{10.29}
$$

The internal energy therefore also determines the temperature.

Finally, let us look at the coefficient of thermal volume change, starting from the first relation (8.15) and using relations (10.27) and (10.26):

$$
\Lambda_V = -T\frac{\partial \hat{S}}{\partial T}\frac{\partial \bar{T}}{\partial V} = -T\frac{1}{\dfrac{\partial \bar{T}}{\partial S}}\frac{\partial \bar{T}}{\partial V} = -T\frac{\bar{C}_V}{\dfrac{\partial \bar{U}}{\partial S}}\frac{\partial \bar{T}}{\partial V} = -\bar{C}_V\frac{\partial \bar{T}}{\partial V} \equiv \bar{\Lambda}_V \tag{10.30}
$$

(we also used relations (10.28) and (10.29)). Referring to relation (10.28) we find that the internal energy determines this coefficient (along with the temperature, which it also ultimately determines). This was the last thing we needed to conclude that the internal energy is also a thermodynamic potential.

And one more nice result for internal energy as a function of volume and entropy. With the help of the second and third relations (7.14), the chain rule, and (10.30) we can write

$$
\begin{aligned}
P = \Lambda_V - \frac{\partial \hat{U}}{\partial V} &= \Lambda_V - \frac{\partial \bar{U}}{\partial V} + \frac{\partial \hat{U}}{\partial T}\frac{\partial \bar{T}}{\partial V} = \Lambda_V - \frac{\partial \bar{U}}{\partial V} + C_V \frac{\partial \bar{T}}{\partial V} \\
&= -C_V \frac{\partial \bar{T}}{\partial V} - \frac{\partial \bar{U}}{\partial V} + C_V \frac{\partial \bar{T}}{\partial V},
\end{aligned}
\tag{10.31}
$$

which ultimately yields the relationship between pressure and internal energy (its partial derivative):

$$
\bar{P} = -\frac{\partial \bar{U}}{\partial V}. \tag{10.32}
$$

The local-global inverse considerations that we performed in deriving the Gibbs energy can be similarly performed here. The inversion condition $\partial \hat{S}/\partial T > 0$ applies locally. It would hold throughout the whole relevant domain if this domain were isochorically convex (a sufficient condition).

10.4 Enthalpy

The Helmholtz and Gibbs energies differ only in the *PV* factor, otherwise there is a subtraction of the *TS* from the internal energy in both. Let us therefore look at the sum $U + PV$, occurring at the Gibbs energy; for the sake of brevity we will denote it by H. We already had the *V*-*T*, *P*-*T* and *V*-*S* combinations as independent variables, we are still missing the *P*-*S* pair. Let us check whether the function *H* would "coincide" with this pair. To introduce the two quantities as independent variables, we have to invert the two (original) functions, $P = \hat{P}(V, T)$ and $S = \hat{S}(V, T)$. If we invert them for volume and temperature, we end up with the new functions $V = \tilde{V}(P, S)$ a $T = \tilde{T}(P, S)$ with the conversion relation:

$$
\hat{f}(V, T) \equiv \hat{f}(\tilde{V}(P, S), \tilde{T}(P, S)) \equiv \tilde{f}(P, S). \tag{10.33}
$$

Can we perform the above inversion? We will verify this by using the appropriate Jacobian, as taught by the mathematics of inverse mappings. Here the situation is somewhat more complicated than in the previous potentials, because we are inverting two functions at the same time. The Jacobian of this pair of functions looks like this:

$$\begin{vmatrix} \dfrac{\partial \hat{P}}{\partial V} & \dfrac{\partial \hat{P}}{\partial T} \\ \dfrac{\partial \hat{S}}{\partial V} & \dfrac{\partial \hat{S}}{\partial T} \end{vmatrix} = \frac{\partial \hat{P}}{\partial V}\frac{\partial \hat{S}}{\partial T} - \frac{\partial \hat{P}}{\partial T}\frac{\partial \hat{S}}{\partial V}. \tag{10.34}$$

If we consider relations (6.6), (6.8), (9.3) and (8.15), we see that this Jacobian is negative. Thus, it is not zero, and the existence of the inversion (i.e., both inverses) implied in (10.33) is guaranteed.[3]

Let us now look at the derivations of the new function under the new pair of variables. We can write the derivative under pressure with respect to (10.33) as follows:

$$\begin{aligned} \frac{\partial \tilde{H}}{\partial P} &= \frac{\partial \tilde{U}}{\partial P} + P\frac{\partial \tilde{V}}{\partial P} + \tilde{V} = \frac{\partial \hat{U}}{\partial V}\frac{\partial \tilde{V}}{\partial P} + \frac{\partial \hat{U}}{\partial T}\frac{\partial \tilde{T}}{\partial P} + P\frac{\partial \tilde{V}}{\partial P} + \tilde{V} \\ &= (\Lambda_V - P)\frac{\partial \tilde{V}}{\partial P} + C_V\frac{\partial \tilde{T}}{\partial P} + P\frac{\partial \tilde{V}}{\partial P} + \tilde{V} = \Lambda_V\frac{\partial \tilde{V}}{\partial P} + C_V\frac{\partial \tilde{T}}{\partial P} + \tilde{V} \\ &= T\frac{\partial \hat{S}}{\partial V}\frac{\partial \tilde{V}}{\partial P} + T\frac{\partial \hat{S}}{\partial T}\frac{\partial \tilde{T}}{\partial P} + \tilde{V}. \end{aligned} \tag{10.35}$$

We have used relations (7.14) and (8.15) to modify the Eq. (10.35). We now use again the familiar relation between partial derivatives of functions of different variables, which we will recall here for convenience by deriving it (again) for entropy, since we will need it for that.

We can write the differential of entropy as a function of the original independent variables (volume, temperature) using the new variables (pressure, entropy) as follows:

$$\begin{aligned} \mathrm{d}S &= \frac{\partial \hat{S}}{\partial V}\mathrm{d}V + \frac{\partial \hat{S}}{\partial T}\mathrm{d}T = \frac{\partial \hat{S}}{\partial V}\Big(\frac{\partial \tilde{V}}{\partial P}\mathrm{d}P + \frac{\partial \tilde{V}}{\partial S}\Big)\mathrm{d}S + \frac{\partial \hat{S}}{\partial T}\Big(\frac{\partial \tilde{T}}{\partial P}\mathrm{d}P + \frac{\partial \tilde{T}}{\partial S}\mathrm{d}S\Big) \\ &= \frac{\partial \hat{S}}{\partial V}\frac{\partial \tilde{V}}{\partial P}\mathrm{d}P + \frac{\partial \hat{S}}{\partial V}\frac{\partial \tilde{V}}{\partial S}\mathrm{d}S + \frac{\partial \hat{S}}{\partial T}\frac{\partial \tilde{T}}{\partial P}\mathrm{d}P + \frac{\partial \hat{S}}{\partial T}\frac{\partial \tilde{T}}{\partial S}\mathrm{d}S \\ &= \Big(\frac{\partial \hat{S}}{\partial V}\frac{\partial \tilde{V}}{\partial S} + \frac{\partial \hat{S}}{\partial T}\frac{\partial \tilde{T}}{\partial S}\Big)\mathrm{d}S + \Big(\frac{\partial \hat{S}}{\partial V}\frac{\partial \tilde{V}}{\partial P} + \frac{\partial \hat{S}}{\partial T}\frac{\partial \tilde{T}}{\partial P}\Big)\mathrm{d}P. \end{aligned} \tag{10.36}$$

By comparing the left and right sides of the last equation, we discover the relationship

$$\frac{\partial \hat{S}}{\partial V}\frac{\partial \tilde{V}}{\partial P} + \frac{\partial \hat{S}}{\partial T}\frac{\partial \tilde{T}}{\partial P} = 0 \tag{10.37}$$

and one more:

$$\frac{\partial \hat{S}}{\partial V}\frac{\partial \tilde{V}}{\partial S} + \frac{\partial \hat{S}}{\partial T}\frac{\partial \tilde{T}}{\partial S} = 1. \tag{10.38}$$

[3] Let us keep in mind that this is again a local proof.

If we substitute the result (10.37) in (10.35), we have

$$\tilde{V} = \frac{\partial \tilde{H}}{\partial P} \tag{10.39}$$

and the function $\tilde{H}$ therefore determines the volume.

And now the derivative by the second variable, entropy:

$$\begin{aligned}\frac{\partial \tilde{H}}{\partial S} &= \frac{\partial \hat{U}}{\partial V}\frac{\partial \tilde{V}}{\partial S} + \frac{\partial \hat{U}}{\partial T}\frac{\partial \tilde{T}}{\partial S} + P\frac{\partial \tilde{V}}{\partial S} = \left(\frac{\partial \hat{U}}{\partial V} + P\right)\frac{\partial \tilde{V}}{\partial S} + \frac{\partial \hat{U}}{\partial T}\frac{\partial \tilde{T}}{\partial S} \\ &= T\frac{\partial \hat{S}}{\partial V}\frac{\partial \tilde{V}}{\partial S} + T\frac{\partial \hat{S}}{\partial T}\frac{\partial \tilde{T}}{\partial S} = T.\end{aligned} \tag{10.40}$$

Here we first used the connection between Eqs. (8.15) and (7.14) and then the result (10.38). We thus found that the function $\tilde{H}$ also determines the temperature:

$$\tilde{T} = \frac{\partial \tilde{H}}{\partial S}. \tag{10.41}$$

Finally, all that remains is to check that the new function also determines Λ_V and C_V. Equations (10.38) and (10.37) can be written as follows with the help of equation:

$$\begin{aligned}\frac{\Lambda_V}{T}\frac{\partial \tilde{V}}{\partial S} + \frac{C_V}{T}\frac{\partial \tilde{T}}{\partial S} &= 1 \\ \frac{\Lambda_V}{T}\frac{\partial \tilde{V}}{\partial P} + \frac{C_V}{T}\frac{\partial \tilde{T}}{\partial P} &= 0,\end{aligned}$$

which is a system of two equations with two unknowns—Λ_V and C_V. The matrix of this system is identical to the (transposed) matrix corresponding to the Jacobian of the pair of functions $\tilde{V}(S, P)$ and $\tilde{T}(S, P)$. As is known from the theory of inverse mappings, this Jacobian is equal to the inverse of the Jacobian given a few lines above and is thus non-zero. Hence, the matrix of our system is regular and the system is solvable. If we substitute in the partial derivatives appearing in it from relations (10.39) and (10.41), its solution will be determined only by the function $\tilde{H}$ (its partial derivatives). Thus this function[4] itself determines $\tilde{\Lambda}_V$ and $\tilde{C}_V$ and is therefore (another) thermodynamic potential. We call it *enthalpy*. With its help, the Gibbs energy can be rewritten to: $G = H - TS$.

[4] The temperature is also determined by this function, see (10.41).

10.5 Summary

Let us summarize the results of our quest for thermodynamic potentials. We have found four, one is the internal energy, and the others are new functions:

$$U = \bar{U}(V, S), \tag{10.42}$$

$$\text{Helmholtz energy: } F = \hat{F}(V, T), \tag{10.43}$$

$$\text{Gibbs energy: } G = \check{G}(P, T), \tag{10.44}$$

$$\text{enthalpy: } H = \tilde{H}(V, S). \tag{10.45}$$

Each has a pair of independent variables for which it acts as a thermodynamic potential. These are then often called the natural variables of the respective quantity (function). Quite generally, we have derived (proved) the potentials only locally in the region where we have defined the entropy, or in the thermodynamic region; under sufficient complementary conditions, we have implied their global existence in such a region.

There are further implications of the properties of thermodynamic potentials as functions which are of interest from a practical point of view. Relations (10.4) and (10.11) imply the following:

$$\frac{\partial \hat{P}}{\partial T} = \frac{\partial \hat{S}}{\partial V}, \tag{10.46}$$

while from relations (10.16) and (10.17) we get

$$\frac{\partial \check{V}}{\partial T} = -\frac{\partial \check{S}}{\partial P}. \tag{10.47}$$

Relations (10.29) and (10.32) imply

$$-\frac{\partial \bar{T}}{\partial V} = \frac{\partial \bar{P}}{\partial S}, \tag{10.48}$$

and finally relations (10.39) and (10.41):

$$\frac{\partial \tilde{T}}{\partial P} = \frac{\partial \tilde{V}}{\partial S}. \tag{10.49}$$

These relations are called *reciprocal relations*. They have practical significance in that the derivatives of (easily) measurable dependencies occur on their left-hand sides. These then allow one to get to the practically immeasurable dependencies, functions and their derivatives on the right-hand sides [4].

This practical approach can be applied when looking at thermodynamic potentials "from the other side". For example, we will then see relation (10.4) as a practical guide to finding the Helmholtz energy as a function of volume—from pressure measurements. We will obtain its dependence on temperature from relation (10.10) using heat capacity and temperature. Measuring the volume will give us the Gibbs energy as a function of pressure, (10.17); similarly, you are sure to find other possibilities.

By combining the first relation of Eq. (7.14) and (9.48) we obtain a relation for the time derivative of the internal energy, and with its help we can easily derive similar relations for the (remaining) thermodynamic potentials:

$$\dot{U} = T\dot{S} - P\dot{V}, \tag{10.50}$$

$$\dot{F} = -P\dot{V} - S\dot{T}, \tag{10.51}$$

$$\dot{H} = T\dot{S} + V\dot{P}, \tag{10.52}$$

$$\dot{G} = V\dot{P} - S\dot{T}. \tag{10.53}$$

By integrating Eq. (10.51) under isothermal conditions in the interval $\langle t_1, t_2 \rangle$ and looking at Eq. (7.2) we find that the energy exchanged by work is equal to the change in Helmholtz energy under these conditions:

$$\Delta F \equiv F(t_2) - F(t_1) = W \quad (T = \text{const.}). \tag{10.54}$$

This is probably why Helmholtz energy is also called free energy. With reference to relation (10.2) it appears as that part of the (internal) energy which is somehow free to do work (under isothermal conditions). For similar reasons, because of the analogous relation $G = H - TS$, Gibbs energy is called free enthalpy.

Equation (10.52) written for constant pressure implies, after substitution from (9.48) and integration, that the change of enthalpy is equal to the energy exchanged by heat $\Delta H = Q_P$ under constant pressure. Many processes in practice are often isobaric, e.g., chemical reactions, therefore when tabulating their thermal effects we find values of enthalpy changes that do not depend on the path of the process. In chemical reactions they are called reaction enthalpies. Similarly, we can use Eq. (10.50) to show the following relation for a constant volume: $\Delta U = Q_V$. Writing the heating in the form of Eq. (6.11), we find, after integrating it under isobaric conditions, that where the inverse conditioned by Eq. (6.8) holds, the analogue of (10.28) applies: $C_P = \partial \check{H} / \partial T$.

10.6 An Overview of the Essentials of Thermodynamics

The complete fundamentals of thermodynamics that we have gone over so far can be summarized in a few points [2]. These are inspired by experience with the thermal behaviour of (especially gaseous) bodies. The fundamental principles are as follows:

1. The simple heating model (6.5) (with coefficients defined on the constitutive domain $\mathcal{D}$).
2. The axiom (6.7) about the existence of pressure as a function of volume and temperature (on the domain $\mathcal{D}$).
3. Axiom I (7.8) on the universal equivalence of energy transfer by heat and work in the cyclic thermal process through which a (fluid) body passes. In it we nowadays choose $J = 1$ by default.
4. Axiom II on the net gain of energy by work (i.e., a positive value $-W$) in ordinary Carnot cycles.
5. Axiom III actually stating that the (value of) work, about which axiom II speaks, is determined only by the operating temperatures of the cycle and the energy absorbed by the heat in it (it does not depend, e.g., on the chemical composition of the body in the cycle).
6. Axiom (8.1) representing Holtzmann's assertion about ideal gases.
7. Axiom on the existence of an ideal gas having as constitutive domain the entire first (positive) quadrant $V - T$.

The first three points allowed us to derive the existence of internal energy (as a function, on the domain $\mathcal{D}$)—see Sect. 7.6. Using axioms II and III, we constructed the corresponding two Carnot cycles, then, in search of limits (see Sect. 9.5) we effectively constructed a sequence of such pairs of cycles, until we were able to show that at any ordinary point lying on a given isotherm the expression (9.1) has the same (nonnegative) value for all fluid bodies. Using the last two points (axioms) we found this value for an ideal gas, and hence (given its proven universality) for all fluids. Then the existence of a function (a quantity; see briefly Sect. 9.1, and more thoroughly again Sect. 9.5), which we called entropy (but which we obtained only on a simply continuous, open subset of the domain $\mathcal{D}$ consisting only of thermodynamic points), was readily apparent. Well, on these foundations we subsequently built the construction of thermodynamic potentials. These seven items are thus quite sufficient to construct the so-called classical thermodynamics, which is also called equilibrium thermodynamics (the reasons for this designation shall remain a mystery for the time being).

Those who do not wish to adhere to the historicist presentation of the essence of thermodynamics can be satisfied with the first three items of this summary, supplemented by the revelation of entropy by means of the universal integration factor, given in Sect. 9.6.

References

1. Partington, J.R.: An Advanced Treatise on Physical Chemistry, vol. 1. The Fundamental Principles - The Properties of Gases. Longmans, London (1949)
2. Truesdell, C., Bharatha, S.:: The Concepts and Logic of Classical Thermodynamics as a Theory of Heat Engines, Rigorously Constructed upon the Foundation Laid by S. Carnot and F. Reech. Springer, New York (1977)
3. Linstrom, P.J., Mallard, W.G. (eds.): NIST Chemistry WebBook, NIST Standard Reference Database Number 69. Gaithersburg: National Institute of Standards and Technology. https://doi.org/10.18434/T4D303 (visited June 6, 2022)
4. Müller, I., Müller, W.H.: Fundamentals of thermodynamics and applications. Berlin, Heidelberg: Springer (2009)

Chapter 11
Entropy Inequality

Abstract In preceding chapters, entropy was obtained for all fluids in the form of an equation relating its time derivative, heating, and temperature. Considering the upper and lower temperature for a process, it is shown mathematically that a more general relationship exists in the form of inequality. Traditionally, it is called the second law of thermodynamics. The relationship between entropy and heat is discussed.

Let us review what we have learned so far about entropy and its time derivative[1]:

$$Q_f = T\dot{S} \quad \Rightarrow \dot{S} = \frac{Q_f}{T} \quad \Rightarrow \Delta S = \int_{t_1}^{t_2} \dot{S}\mathrm{d}t = \int_{t_1}^{t_2} \frac{Q_f}{T}\mathrm{d}t. \tag{11.1}$$

We can play a little mathematical game with integration in these relationships. First, let us point out one important thing. For the Carnot cycle, the body was between two extreme (operating) temperatures—the minimum and maximum ones. This may not be a speciality of this cycle, apparently for any thermal behaviour of a body, its temperature in the relevant process is between some upper and lower limit:

$$0 < T_{\min} \le T(t) \le T_{\max}, \quad t \in \langle t_1; t_2 \rangle \tag{11.2}$$

(the function $\check{T}(t)$ is positive and continuous on that interval). Then, taking into account the notation of the heating integrals given in Sect. 9.7.2 (where we substitute the inverse of temperature for f), we can write the following inequalities [1]:

$$\begin{aligned} \int_{T^+} \frac{Q_f}{T}\mathrm{d}t &= \frac{1}{2}\int_{t_1}^{t_2} \frac{1}{T}(|Q_f| + Q_f)\mathrm{d}t \ge \frac{Q^+}{T_{max}}, \\ -\int_{T^-} \frac{Q_f}{T}\mathrm{d}t &= \frac{1}{2}\int_{t_1}^{t_2} \frac{1}{T}(|Q_f| - Q_f)\mathrm{d}t \le \frac{Q^-}{T_{min}}. \end{aligned} \tag{11.3}$$

[1] Let us recall that heating is assumed to be an integrable function and this is closely related to the continuity of such a function—see Sect. 6.2.

© The Author(s), under exclusive license to Springer Nature Switzerland AG 2024

M. Pekař, *The Essentials of Thermodynamics*, Physical Chemistry in Action,
https://doi.org/10.1007/978-3-031-60321-1_11

Subtracting them and using (11.1), we obtain

$$\frac{Q^+}{T_{max}} - \frac{Q^-}{T_{min}} \le \int_{t_1}^{t_2} \frac{Q_f}{T} dt = \Delta S. \tag{11.4}$$

The inequality (11.4) allows us to derive another interesting and more general consequence [2]. The change referred to by the symbol Δ takes place in a certain time interval $\langle t_1; t_2 \rangle$, which for further simplification we denote by $\langle t_0 - h; t_0 + h \rangle$ (the point t_0 thus lies in the middle of the interval $\langle t_1; t_2 \rangle$). We generalize equation (11.4) by introducing the parameters a and b instead of the maximum and minimum temperatures, and we express the energy exchanged by heat absorbed and emitted occurring in this equation using their definitions (6.3) and (6.4):

$$\begin{aligned} \frac{Q^+}{a} - \frac{Q^-}{b} &= \frac{1}{2b} \int_{t_0-h}^{t_0+h} (Q_f + |Q_f|) \mathrm{d}t + \frac{1}{2a} \int_{t_0-h}^{t_0+h} (Q_f - |Q_f|) \mathrm{d}t \\ &= \frac{1}{2}\Big(\frac{1}{a} + \frac{1}{b}\Big) \int_{t_0-h}^{t_0+h} Q_f \mathrm{d}t - \frac{1}{2}\Big(\frac{1}{a} - \frac{1}{b}\Big) \int_{t_0-h}^{t_0+h} |Q_f| \mathrm{d}t. \end{aligned} \tag{11.5}$$

Here, a and b are positive parameters related to (and constant over) the given time interval. If we choose a fixed t_0, the values of these parameters are a function only of the "width" of the interval h. Since we assume the continuity of Q_f on the interval $\langle t_0 - h; t_0 + h \rangle$, for a given choice of t_0, we can express the above integrals using the mean value theorem,[2] which operates on the interval $\langle -h; h \rangle$ and with the numbers c and d of that interval:

$$\frac{1}{2h}\Big(\frac{Q^+}{a} - \frac{Q^-}{b}\Big) = \frac{1}{2}\Big(\frac{1}{a} + \frac{1}{b}\Big) Q_f(t_0 + c) - \frac{1}{2}\Big(\frac{1}{a} - \frac{1}{b}\Big) |Q_f(t_0 + d)|. \tag{11.6}$$

With the help of (11.6) the relation (11.4), generalized to the form

$$\Delta S \ge (Q^+/b) - (Q^-/a), \tag{11.7}$$

can be rewritten:

$$\frac{\Delta S}{2h} \ge \frac{1}{2}\Big(\frac{1}{a} + \frac{1}{b}\Big) Q_f(t_0 + c) - \frac{1}{2}\Big(\frac{1}{a} - \frac{1}{b}\Big) |Q_f(t_0 + d)|. \tag{11.8}$$

Let us further assume that the function $\hat{S}$ is differentiable at t_0[3] and that both a and b are continuous functions of h at $h = 0$. Combining the differentiability and relation (11.8), we then successively derive

---

[2] If $\hat{f}(x)$ is continuous in $\langle a; b \rangle$, then there exists $c \in \langle a; b \rangle$ such that $\int_a^b \hat{f}(x) \mathrm{d}x = (b - a) f(c)$.

[3] A function $\hat{f}(x)$ is differentiable at a if for its increment: $\Delta \hat{f} \equiv f(a + h) - f(h) = f'(a)h + h\hat{\tau}(h)$, where $\lim_{h \to 0} \hat{\tau}(h) = 0$.

$$\frac{\Delta S}{2h} = \dot{S}(t_0) + \hat{\tau}(h) \geq \frac{1}{2}\Big(\frac{1}{a} + \frac{1}{b}\Big) Q_f(t_0 + c) - \frac{1}{2}\Big(\frac{1}{a} - \frac{1}{b}\Big)|Q_f(t_0 + d)|,$$

$$\dot{S}(t_0) \geq \frac{1}{2}\Big(\frac{1}{a} + \frac{1}{b}\Big) Q_f(t_0 + c) - \frac{1}{2}\Big(\frac{1}{a} - \frac{1}{b}\Big)|Q_f(t_0 + d)| - \hat{\tau}(h)$$

and in the limit $h \to 0$:

$$\dot{S}(t_0) \geq \frac{1}{2}\Big(\frac{1}{a(0)} + \frac{1}{b(0)}\Big) Q_f(t_0) - \frac{1}{2}\Big(\frac{1}{a(0)} - \frac{1}{b(0)}\Big)|Q_f(t_0)| \tag{11.9}$$

The inequality (11.9) then implies the following for different values of heating:

$$\dot{S}(t_0) \geq \begin{cases} \dfrac{Q_f(t_0)}{b(0)} & \text{for} \quad Q_f \geq 0, \\ \dfrac{Q_f(t_0)}{a(0)} & \text{for} \quad Q_f \leq 0. \end{cases} \tag{11.10}$$

Finally, if we choose $a = T_{min}$, $b = T_{max}$, then $\lim_{h\to 0} \hat{a}(h) = \lim_{h \to 0} \hat{b}(h) = T(t_0)$, and from (11.10) we get (locally) the inequality

$$T\dot{S} \geq Q_f \tag{11.11}$$

and by its integration[4]

$$\Delta S \geq \int_{t_1}^{t_2} \frac{Q_f}{T} \mathrm{d}t. \tag{11.12}$$

What did we actually derive? Instead of the equalities appearing in equations (11.1), we have the related inequalities (11.11) and (11.12). Isn't that a kind of contradiction? These inequalities alert us that the equation appearing in Eq. (11.4) was too strict; it was a consequence of our special model for heating (6.5), which we actually substituted into (11.4) at the end. If we require satisfying "only" inequality (11.7), where we subsequently choose the boundary temperatures of the processes as parameters a, b, satisfying the weaker conditions (11.11) and (11.12) is sufficient.

The relation (11.11) is called the *second law* of thermodynamics or also *entropy inequality*.

Our heating model is very simple indeed. It assumes that the exchange of energy by heat in the modelled body occurs instantaneously and uniformly throughout the body. The body is always at the same temperature everywhere and the change in volume is also uniform, there are no local irregularities. The heat exchange is only manifested by a change in temperature, that is, roughly by a change in the kinetic energy of the particles of the body, or by a change in its volume. The change in

[4] Integrability is roughly based on the following. For heating and temperature as functions of time, we have introduced continuity assumptions. We know from calculus that if the functions $\hat{f}(x)$ and $\hat{g}(x)$ are continuous at c, then under the condition of non-zero $g(c)$, the function $\hat{f}(x)/\hat{g}(x)$ is also continuous at c [3].

volume can be attributed to the change in the state of motion of the particles related to temperature—the changed speed of the particle motion is also related to the change in the size of the space in which the motion takes place. We can look at the same temperature everywhere when heating a body from two seemingly contradictory points of view. Either so that the exchange of energy by heat is so (infinitely) slow that the temperature still manages to equilibrate throughout the volume of the body. Or in such a way that the energy exchanged by heat is immediately (infinitely fast) evenly "distributed" over the whole body. In reality, heating does not simply occur in this way. The heating of a well-mixed liquid in a container could approximately correspond to our model, if we do not take into account the motion of the whole liquid due to the mixing. In reality, however, we are much more often dealing with the distribution of, say, temperature in a given body—the temperature field; this is beyond the scope of this text.

Also, in reality, heating does not necessarily lead to a change in temperature or volume; this is of course a kind of "lossless" idealization of energy exchange, which is in fact just a (uniform) change in the kinetic energy of the particles of the body. The exchanged energy is not "lost" in our model to anything else, to some other internal phenomena in the body. Thus, we can reasonably assume that outside our heating model, the (more general) inequality (11.11) or (11.12) will apply to entropy.

We can now also generalize Eq. (9.64) or the relationship between entropy and heat

$$Q = \int_{t_0}^{t} Q_f dt \leq [TS]_{t_0}^{t} - \int_{t_0}^{t} \dot{T} S dt. \tag{11.13}$$

Truesdell explains the entropy inequality in an untypical way—as the existence of an upper bound on the rate at which a body can exchange (internal) energy with heat, rather than exchanging it with work [4]. This bound is thus the product $T\dot{S}$, cf. relation (11.11), having the dimension energy per time.

References

1. Truesdell, C.: The efficiency of a homogeneous heat engine. J. Math. Phys. Sci. **7**(4), 349–371 (1973); **9**(2), 198–199 (1975)
2. Truesdell, C.: Proof that my work estimate implies the Clausius-Planck inequality. Rendiconti della classe di scienze fisiche, matematiche e naturali **68**(8), 191–199 (1980)
3. Jarník, V.: Differential Calculus I. Academia, Praha (1974). (in Czech)
4. Truesdell, C.: Rational Thermodynamics, 2nd edn. Springer, New York (1984)

Chapter 12
Another Generalization, This Time Concerning Entropy Inequality

Abstract The notion of dissipation is examined, two meanings are given, and a quantity called dissipation is introduced. The first and second laws are generalized for uniform bodies. Reduced entropy inequality is formulated as a general principle to be fulfilled by any body and its constitutive equations. As an example, all the thus-far-exposed-structure of thermodynamics is derived from this principle. Equilibrium states and processes are addressed. A more complex constitutive model is shown, which enables also non-equilibrium processes. Thermodynamic potentials and the meaning of entropy under the expounded generalization are dealt with.

12.1 On Dissipation

The notion of *dissipation*, more precisely energy dissipation, often revolves around entropy. The word, of Latin origin, has the meaning of dispersal, dispersion of energy. The term dissipation was probably first used by Kelvin [1] in the sense of the conversion of mechanical energy into a kind of waste energy that cannot be used much (for work). He says "when heat is created by any irreversible process (such as friction), there is a dissipation of mechanical energy and a full restoration of it to its primitive condition is impossible". Truesdell writes that the concept of dissipation was introduced by Kelvin in a qualitative discussion of natural phenomena in which mechanical energy is lost [2]. Since then it has somehow expressed the "conversion" of energy and mechanical work into heat. For example, Owen sees dissipation somewhat differently [3]: viscous forces (a kind of internal friction that can occur in many bodies) in a homogeneous body cannot improve the efficiency of a heat machine that uses such a body as a working substance, rather the opposite. Therefore, viscous forces are called dissipative, in the sense of waste, leading to waste.

Dissipation can thus be understood in a double meaning. Firstly, as a kind of loss of energy by heat transfer in the process of energy exchange between bodies (i.e., a non-zero value of Q^-). Secondly, in a material sense, as the existence of forces within a body which prevent the full conversion of energy (supplied by heat, internal energy, etc.) into work. The two meanings are certainly intertwined.

© The Author(s), under exclusive license to Springer Nature Switzerland AG 2024
M. Pekař, *The Essentials of Thermodynamics*, Physical Chemistry in Action,
https://doi.org/10.1007/978-3-031-60321-1_12

However, entropy as a measure or expression of dissipation must be treated with caution. In the first sense, we should have a clear relation between entropy and emitted heat. We can derive such a relationship from the Eq. $(11.3)_2$ for a part of the cycle:

$$-T_{min}\int_{T^-}\dot{S}\mathrm{d}t \leq Q^-; \tag{12.1}$$

it indicates the non-zero value of Q^- (of "heat loss") due to the non-zero entropy (its time derivative). More precisely, it shows that the losses will be zero if the entropy is invariant and under only the equal sign in (12.1). But this is a rather empty or tautological solution given the definition equation for entropy, e.g., (9.48). If the entropy does not change, then by (9.48) the heating is also zero and there is no point in talking about losses. We reach a similar conclusion if we understand the first sense as the absence of losses when entropy is zero. The same follows from the relationship between total heat and entropy, e.g., (9.64). On the other hand, in Sect. 9.8 we saw lossless heat-work conversion without entropy being zero (or constant) in the isothermal behaviour of an ideal gas. Furthermore, in Sect. 9.7.3 we revealed the inevitability of losses in a cycle in which the entropy change is zero, however. Let us recall, though, that this inevitability was closely related to the existence of an integration factor, which was also the "cause" of the existence of entropy.

Thus, the first, loss-related meaning of entropy as a measure of dissipation seems quite problematic. Let us examine the second, material-related one. Our initial material, is non-dissipative, as already mentioned. The energy supplied to it by heat goes only to a change in temperature or volume, no "internal friction" or similar dissipative phenomena. However, even here losses cannot be avoided, see again Sect. 9.7.3. The difference with more complex materials, however, is precisely in the entropy—our basic material has an equal sign in relation (11.11), "dissipative" materials have (also) a non-equal sign. The inequality is precisely because the energy exchanged by heat can be "lost" (not literally, due to conservation) to dissipative phenomena, like that internal friction, and then the heating divided by the corresponding temperature is not equal to the time change in entropy but is smaller. We can then understand that this diminishing just characterizes, perhaps even directly measures, dissipation. Without dissipative diminishing $T\dot{S} - Q_f = 0$, so we could interpret this result as zero dissipation (energy in the body) or at least a consequence of zero dissipation. For these reasons, the quantity or product $T\Sigma$, where Σ is defined in this way

$$\Sigma = \dot{S} - \frac{Q_f}{T}, \tag{12.2}$$

is usually directly called *dissipation*. When it is zero, there is obviously an equation in the entropy inequality. Otherwise it can only be positive according to this inequality.

Dissipation as a concept and as a quantity ($T\Sigma$) is intended to express the conversion of energy and mechanical work "into heat". According to the first law, if energy is not exchanged by heat, all work goes into changing internal energy, entropy is invariant and $T\Sigma$ zero in a non-dissipating body (equation in the second law), whereas in a dissipating body (non-equation in the second law) entropy increases over time and dissipation $T\Sigma$ is positive. If energy is exchanged by heat, the key is the division of heat into absorbed and emitted, only the latter representing the dissipation of energy. In a dissipating body the dissipation of $T\Sigma$ is also positive when energy is exchanged by heat, in a non-dissipating body it is again zero. Substituting into the definition of dissipation from the first law, we can see that it expresses the difference between the time derivative of entropy and the change in internal energy not caused by work (not caused by the power of external forces, given by the difference $(\dot{U} - (-P\dot{V}))$: $T\Sigma = \dot{S} - (\dot{U} + P\dot{V})$. If the dissipation is zero, this difference is also zero; otherwise the dissipation and the difference can only be positive.

The dissipation and its measure associated with entropy will thus be understood here in the sense of the product quantity $T\Sigma$, i.e., rather in the latter, material sense.

The symbol Σ is sometimes called entropy production. This is a misleading name giving the impression that entropy is something that can be produced, so we will not use it here.

12.2 Generalized First and Second Law

With our basic heating model, we have been gradually reaching interesting results and discovering new quantities on our way to essentials of thermodynamics. Let's try to look for general elements along this way. The internal energy is certainly a general quantity due to the general atomic-molecular structure of bodies, it can vary in time, so it is fundamentally a function of time.

The first law, as a law of conservation of energy, is also a general principle (we can safely say axiom). In general, when we do not have to consider changes in other kinds of energy, we can write it in a form determining the time derivative of the internal energy $\dot{U} = Q_f + W_f$, where W_f denotes the (work) power of the forces acting and has the energy/time dimension. If the only force acting is the pressure force, which we will restrict ourselves to here, $W_f = -P\dot{V}$ (pressure and volume are generally functions of time, as is temperature). Heating describes the transfer of energy caused by the difference in temperature of the bodies between which the energy is transferred. The power (work) then expresses the transfer of energy given by the action of a force, a mechanical transfer, here limited to the transfer by the action of a pressure force on a given body.

Let us also declare the second law in the form of an entropy inequality to be a general principle which has universal validity like the conservation of energy. This also postulates the existence of entropy (as a function of time).

Finally, we have considered all (thermodynamic) quantities as functions of two fundamental quantities, volume and temperature, which have thus replaced time as the independent variable. However, this will no longer be a universally valid principle, but rather our simple model. These functions actually represent constitutive (state) equations as models of a body. There may be many models as well as bodies, but these functions, the constitutive equations, must satisfy the two general laws.

So let's see what this (general) requirement means for (particular) functions. Let's mention that we will be concerned with general mathematical laws and derivations, we don't have to define the constitutive domain a priori and, for example, for temperature and volume to consider quite generally, without further emphasis, the whole first (positive) quadrant. The validity of the derived relations (of the most general model) can be limited to a narrower region when describing a particular body, and its determination is then a question of this particular application of the general model.

Let's combine the two laws into one expression:

$$-T\Sigma = -T\dot{S} + \dot{U} + P\dot{V} \leq 0, \tag{12.3}$$

which is called the *reduced entropy inequality*.

12.3 Generalization of Our Macro-Model

Let us substitute into Eq. (12.3) the time derivatives of entropy and internal energy resulting from their nature as functions of volume and temperature. We leave the time derivative of volume as the independent variable (one of the two independent variables) unchanged. After a simple adjustment, we obtain the inequality:

$$-T\Sigma = \Big(-T\frac{\partial \hat{S}}{\partial V} + \frac{\partial \hat{U}}{\partial V} + \hat{P}(V,T)\Big)\dot{V} + \Big(-T\frac{\partial \hat{S}}{\partial T} + \frac{\partial \hat{U}}{\partial T}\Big)\dot{T} \leq 0. \tag{12.4}$$

Volume and temperature are independent variables, they can take (at least mathematically) arbitrary values, and we can choose the time derivatives of volume and temperature arbitrarily, since we have no mathematical constraints on them, but inequality (12.4). must always be satisfied. This situation is close to a fairly simple mathematical statement.

That statement concerns the inequality $ax + b \leq 0$, where a and b are (real) constants, which is supposed to apply to arbitrary real values of x. The statement states that a necessary and sufficient condition for this is that $a = 0$ and $b \leq 0$. Otherwise, it would be impossible to find x, for which inequality would not apply, as one can easily see. Let us call this statement the *inequality theorem*.

We can convert our inequality (12.4) to the case of the above inequality in two ways. First: let us choose arbitrary but fixed values for the independent variables V and T and of the derivative $\dot{T}$; the derivative $\dot{V}$ remains arbitrary, thus fulfilling the role of x in the inequality of the previous paragraph. Second: let us choose arbitrary but fixed values of the independent variables V and T and—this time—of the derivative $\dot{V}$; x is now represented by the (arbitrary value of) the derivative of $\dot{T}$. In the first case, we get the following equality from the inequality theorem

$$\frac{\partial \hat{U}}{\partial V} + P = T\frac{\partial \hat{S}}{\partial V} \tag{12.5}$$

and in the second case the equality

$$\frac{\partial \hat{U}}{\partial T} = T \frac{\partial \hat{S}}{\partial T}. \tag{12.6}$$

Although we have chosen V and T fixed, they are completely arbitrary, so that the obtained equations are valid for all values of volume and temperature. The first equality results from the combination of the first Eq. (8.15) and the first equality in Eq. (10.31). The second equality is the same as Eq. (10.23). From these two equalities, we can derive all the other relationships we have derived so far for thermodynamic quantities (functions) by interconverting between the functions in a similar way as in the previous chapters. This means that our generalization works! From just two laws (axioms) and the detection of the independent variables in constitutive equations, we can derive the entire mathematical structure of macroscopic thermodynamics that we have shown so far.

Let us also note that we have shown a sort of pure mathematical approach, in derivation we seem to choose arbitrarily the values of volume or temperature. The applicability of a finite model, satisfying both axioms, may then ultimately be limited in practical application to some subset of its actual constitutive domain. Thus, it need not describe the behaviour of a given body under all conditions. Just as the ideal gas model describes the behaviour of real gases only in a certain range of temperatures and pressures.

The sharp-minded reader may have noticed that the fewer time derivatives of thermodynamic quantities we have in the reduced entropy inequality, the simpler the further derivation. Let us further illustrate this by considering the case where we substitute the free (Helmholtz) energy (10.2) from the definition of free energy for the internal energy. The reduced entropy inequality then reads

$$-T\Sigma = \dot{F} + P\dot{V} + S\dot{T} \leq 0. \tag{12.7}$$

By substituting the time derivative of the free energy and a simple modification we finally get

$$-T\Sigma = \left(\frac{\partial \hat{F}}{\partial V} + \hat{P}(V,T)\right)\dot{V} + \left(\frac{\partial \hat{F}}{\partial T} + \hat{S}(V,T)\right)\dot{T} \leq 0. \tag{12.8}$$

By a fixed but arbitrary choice of T, V, and $\dot{V}$ we obtain the relation (10.11) from the inequality theorem in which $\dot{T}$ plays the role of the variable x. By a fixed but arbitrary choice of T, V, and $\dot{T}$ we obtain the relation (10.4) from the inequality theorem in which $\dot{V}$ plays the role of the variable x.

12.4 Equilibrium Processes and States

Let us notice one more important point, which is actually a consequence of the first generalization presented in this chapter. The consequence of relations (12.5) and (10.23) or (10.11) and (10.4) is always just an equation in the (reduced) entropy inequality. Let us name the processes in which the equality sign in the entropy inequality applies (whether "original" or reduced), the *equilibrium processes*. Thus, only equilibrium processes are possible in our model. By the model we mean a body the properties of which are determined by volume and temperature, with the first law of the form $\dot{U} = Q_f - P\dot{V}$ and the second law of the form $\dot{S} = Q_f/T$; the heating model (6.5) is not relevant in the generalization of this chapter. Moreover, the body (in this model) has the same temperature everywhere and in general—it has no spatial inhomogeneities, its properties change at most with time. We call such a body a *uniform body*.

Let us explain the choice of the name equilibrium process. We start with the related term equilibrium state. Intuitively, under the term *equilibrium state* of a body we understand such its state, when in time "nothing happens", its properties do not change, it is in equilibrium with neighbouring bodies. The time derivatives of volume, temperature and other thermodynamic quantities are zero, there is no exchange of energy by heat or work. In the second law, the equation (only) applies, because both of its terms ($\dot{S}$ and Q_f) are zero. So the dissipation (and, certainly, Σ) is zero, because both terms that make up the difference are zero. This is not the only case where the dissipation can be zero; the other is the case where $\dot{S} = Q_f/T$, and the time derivatives are not zero. So something is going on in time, but in such a "balanced" way that the dissipation is zero, just as in the equilibrium state. Again, then only the equality sign in the second law applies. Because something is happening in time, we call the relevant processes equilibrium processes. So the concept of equilibrium process in terms of zero dissipation generalizes the concept of equilibrium state. The equilibrium state is then a special case of the equilibrium process, special in the sense that in the given state the body still remains.

We will still look at equilibrium processes through the lens of thermodynamic potentials, popular in the traditional interpretation of basic thermodynamics. By looking at them through their natural variables. Consider Eq. (12.7) in the case where the natural variables of the Helmholtz energy (volume and temperature) are invariant. It simplifies to the form $\dot{F} \leq 0$. Or rather, in the equilibrium, $\dot{F} = 0$. This means that in equilibrium at constant volume and temperature the Helmholtz energy is invariant. Out of equilibrium, however, it could only decrease ($\dot{F} < 0$). Thus, we get one of the criteria of the direction of the course of processes and equilibrium: at constant volume and temperature the processes proceed in the direction of the decrease of Helmholtz energy until the equilibrium, when the Helmholtz energy is minimum and constant.

Dear readers, put the Gibbs energy instead of the Helmholtz energy in Eq. (12.7) and derive a similar criterion for the case of invariant natural variables of the Gibbs energy: at constant pressure and temperature the processes proceed in the direction of the decrease of the Gibbs energy, until the equilibrium, when the Gibbs energy is minimum and constant. It's a very simple exercise. Let us just recall, that in the model of the body just discussed, only equilibrium processes are possible, as we stated at the beginning of this section. This is also clear from the following discussion. If the Helmholtz energy is a function of volume and temperature only, and these quantities are constant, then the Helmholtz energy is also constant and only the equality sign in relation (12.7) applies.

12.5 A More Complex Macro-Model

Let us see where this generalization leads if we choose a slightly more complex model. Just the two simple independent variables used so far have meant that the (thermodynamic) properties of a body at any moment are determined only by the values of these variables at that moment, independent of the previous history of the body. Let us add another independent variable to these, namely the time derivative of temperature ($\dot{T}$). The derivative generally describes the shape of the function at the point at which the derivative is determined, showing its steepness, even in a certain neighbourhood of that point. The time derivative of temperature characterizes the rate of change of temperature and thus to some extent describes the temperature history of a body. If the time derivative is small at a given time, the temperature had changed little (slowly) before that point and vice versa. By including $\dot{T}$ among the independent variables, we say that the properties of the body at a given point are also determined by its (not too distant) temperature history; we are thus modelling a body with some memory with respect to temperature. Otherwise the body remains uniform, without any spatial inhomogeneities.

We start from the reduced inequality (12.7), but the Helmholtz energy will be a different function this time: $\hat{F}(V, T, \dot{T})$. Substituting this function into (12.7) after a small modification we get

$$-T\Sigma = \left(\frac{\partial \hat{F}}{\partial V} + \hat{P}(V, T, \dot{T})\right)\dot{V} + \left(\frac{\partial \hat{F}}{\partial T} + \hat{S}(V, T, \dot{T})\right)\dot{T} + \frac{\partial \hat{F}}{\partial \dot{T}}\ddot{T} \leq 0. \quad (12.9)$$

We can freely choose the values of the independent variables and their (first or second) time derivatives. If we choose arbitrary but fixed values of V, T and $\dot{T}$, the role of the variable x of the inequality theorem remains to be played by $\ddot{T}$. Then it follows:

$$\frac{\partial \hat{F}}{\partial \dot{T}} = 0. \quad (12.10)$$

This, in turn, applies to all values of the three independently chosen variables, but this means that the Helmholtz energy is ultimately not a function of the time derivative of the temperature, and remains the same function as in the previous model:

$$F = \hat{F}(V, T). \tag{12.11}$$

If we choose the independent variables and the time derivatives except $\dot{V}$ arbitrary but fixed, then the role of the variable x plays just the $\dot{V}$ and from inequality (12.9) we get the equality

$$\frac{\partial \hat{F}}{\partial V} = -P. \tag{12.12}$$

However, given (12.11) it follows that the pressure is also a function of volume and temperature only, $P = \hat{P}(V, T)$.

We can get no more from inequality (12.9). For we cannot arrange for the derivative $\dot{T}$ to play the role of the (arbitrarily varying, remember) variable x—because the value of entropy depends on it (since we have not got rid of the dependence on $\dot{T}$) and then we would not have a constant for this variable as required by the inequality theorem (the constant a). So in the end, we are left with the entropy inequality in the shorter form for this model:

$$-T\Sigma = \left(\frac{\partial \hat{F}(V, T)}{\partial T} + \hat{S}(V, T, \dot{T})\right)\dot{T} \leq 0. \tag{12.13}$$

Thus, obviously, in this model, in the entropy inequality (or for dissipation), the equality sign does not always apply, but we are dealing with a model in which non-equilibrium states also occur. These must satisfy relation (12.13), in which there is only an inequality. Let us then define naturally the equilibrium process, briefly equilibrium, in this body as the process when the following applies: $\dot{T} = 0$. Naturally, because this is the condition (for the independent variable) that makes both Σ and dissipation equal to zero, as we require for equilibrium.

It follows from the result (12.13) that the dissipation, and hence Σ, can only be positive or zero; therefore, its zero value is the minimum value. In other words, Σ reaches its minimum in equilibrium. The mathematics for the minimum of a function specifies clear conditions. Let us examine their implications for our model under discussion. Which function will we be concerned with? The function $\hat{\Sigma}(V, T, \dot{T})$. The equilibrium is defined and determined by the zero time derivative of temperature, i.e., only one of the three independent variables, so the other two do not "interest" us in the discussion of the minimum, they have no effect on it. Therefore, we will use a little mathematical trick and examine the minimum of the function of the (real) parameter λ: $\hat{\Sigma}(V, T, \lambda\dot{T})$, which occurs when $\lambda = 0$. The minimum condition (for arbitrarily, though fixed, chosen V, T, $\dot{T}$) is then:

$$\left.\frac{d\hat{\Sigma}(V, T, \lambda\dot{T})}{d\lambda}\right|_{\lambda=0} = 0, \tag{12.14}$$

$$\left.\frac{d^2\hat{\Sigma}(V, T, \lambda\dot{T})}{d\lambda^2}\right|_{\lambda=0} \geq 0. \tag{12.15}$$

Performing the first derivative, we get

$$\frac{d\hat{\Sigma}}{d\lambda}\Big|_{\lambda=0} = -(1/T)\Big[\frac{\partial \hat{F}(V,T)}{\partial T} + \hat{S}(V,T,0)\Big]\dot{T} = 0. \tag{12.16}$$

This equation must be satisfied for arbitrary values of $\dot{T}$ (for each equilibrium), which requires the following equality:

$$\frac{\partial \hat{F}(V,T)}{\partial T} = -\hat{S}(V,T,0) \equiv -\hat{S}_{\text{eq}}(V,T), \tag{12.17}$$

where we have introduced an equilibrium entropy S_{eq}, depending only on volume and temperature. The relation (12.17) applies to all values of volume and temperature, since these were chosen arbitrarily in the derivation. The entropy can thus be expressed as the sum of the equilibrium and non-equilibrium entropy:

$$\hat{S}(V,T,\dot{T}) = \hat{S}_{\text{eq}}(V,T) + \hat{S}_{\text{neq}}(V,T,\dot{T}). \tag{12.18}$$

The non-equilibrium entropy is of course zero in equilibrium, as expressed by the following additional condition:

$$\hat{S}_{\text{neq}}(V,T,0) = 0. \tag{12.19}$$

Substituting the definition (12.18) and the condition (12.19) into the original entropy inequality (12.13), we obtain (with the consideration (12.16)) its modified form:

$$\hat{S}_{\text{neq}}(V,T,\dot{T})\,\dot{T} \le 0. \tag{12.20}$$

Performing the second derivative we get

$$\frac{d^2\hat{\Sigma}}{d\lambda^2}\Big|_{\lambda=0} = (2/T)\dot{T}^2\frac{\partial \hat{S}(V,T,0)}{\partial \dot{T}} \le 0. \tag{12.21}$$

Satisfying this inequality requires that

$$\frac{\partial \hat{S}(V,T,0)}{\partial \dot{T}} \equiv \frac{\partial \hat{S}_{\text{neq}}(V,T,0)}{\partial \dot{T}} \le 0. \tag{12.22}$$

In the identity Eq. (12.22), we have used the entropy decomposition (12.18).

Let us summarize the implications of our generalized procedure for that more complex model. Here too, Helmholtz energy and pressure are functions of volume and temperature only. Non-equilibrium processes are possible in it, which must satisfy conditions (12.13), or (12.20), or (12.22). The non-equilibrium processes are characterized by the non-equilibrium entropy S_{neq}, which occurs in the last two conditions.[1] Using relations (12.12) and (12.17), we can write for the time derivative of the free energy:

$$\dot{F} = -P\dot{V} - S_{\text{eq}}\dot{T}. \tag{12.23}$$

This is an example of the so-called *Gibbs equation* (we have seen others before cf. (10.50)–(10.53)).

12.6 Generalizations and Thermodynamic Potentials

Equations (12.12) and (12.17) are quite similar to Eqs. (10.4) and (10.11), which occurred when we encountered thermodynamic potentials. Except that, by the definition of the term, these were also supposed to determine the two coefficients in the heating model. Here we no longer have a special heating model. However, at least one of those coefficients, the heat capacity, is a commonly established quantity in the study of thermal phenomena. Let's see if we can learn something about it from the generalization discussed.

We have heating captured (only) in the first law in general. For the more complex model discussed in the previous section, we can express from it as follows:

$$\begin{aligned} Q_f &= \frac{\partial \hat{U}}{\partial V}\dot{V} + \frac{\partial \hat{U}}{\partial T}\dot{T} + \frac{\partial \hat{U}}{\partial \dot{T}}\ddot{T} + P\dot{V} \\ &= \left(\frac{\partial \hat{U}}{\partial V} - \frac{\partial \hat{F}}{\partial V}\right)\dot{V} + \frac{\partial \hat{U}}{\partial T}\dot{T} + \frac{\partial \hat{U}}{\partial \dot{T}}\ddot{T}. \end{aligned} \tag{12.24}$$

The coefficient at $\dot{T}$ was the heat capacity in the original heating model. What it looks like here in the last equation is perhaps even the same as in Eq. (10.28). But beware, here we have internal energy as a different function, and moreover, as a potential, the free rather than the internal energy appears here—given Eqs. (12.12) and (12.17). We resolve this by relaxing for a moment the rigour of defining what we mean by the thermodynamic potential, and by immediately defining the heat capacity C_V as is customary in classical thermodynamics, as the partial derivative of the internal energy with respect to temperature; thus we take Eq. (10.28) as a definition. This gives us the C_V also in the heating equation (12.24), but note that here, unlike Eq. (10.28), it is a function of volume, temperature, and even $\dot{T}$ (the same as the internal energy)! So the definition is seemingly the same, but the relevant function is different, and then even in the experimental determination for a body that would follow the more complicated model, we must be careful to observe the conditions of the partial derivative definition (i.e., the C_V is determined here by the definition under constant volume and constant $\dot{T}$). In other words, the mathematical definition of the heat capacity C_V as the partial derivative of the internal energy with

[1] One would like to say that it is their cause, but that would be too forced, the cause does not directly follow here, only the characteristic, the quantity describing the non-equilibrium processes.

temperature remains, we just have to keep in mind what function we are partially deriving. For this more complex model, its experimental determination would then mean holding (in addition to volume) the change in temperature over time constant, which is satisfied, for example, if temperature varies linearly with time: $T = at + b$ (a and b are real constants).

The coefficient at $\dot{V}$ was the coefficient of thermal volume change Λ_V. Here, in a more complex model, it could therefore be defined as

$$\Lambda_V = \frac{\partial \hat{U}}{\partial V} - \frac{\partial \hat{F}}{\partial V}. \tag{12.25}$$

But then again, it is not only determined by free energy. Let us then weaken our definition of thermodynamic potential and call it in general a function which, by its derivatives, determines some thermodynamic quantity. In the case of the more complex model under discussion, then, free (Helmholtz) energy (as a function of volume and temperature) is the potential for pressure and (only) equilibrium entropy. This is ultimately very close to its role in the basic model, for which only equilibrium processes could occur; cf. just Eqs. (10.4) and (10.11).

12.7 Further to the Meaning of Entropy

When we first encountered entropy, in Chap. 9, we noted that entropy is a function; one of the functions we use to describe our thermal experience. We did not have a clear definition, form, or formula for it; we obtained a clear equation (only) for its time derivative. In this chapter we seem to be even worse off with entropy; we no longer have a clear equation even for this derivative of it. We have "only" an inequality for this derivative. But no matter, we have shown how and that with the help of the general functional expression of entropy and with the help of other thermodynamic quantities—functions (potentials) we can derive a number of useful relations and also conditions for non-equilibrium processes. Some of these relations at least show us the relation of the partial derivatives of entropy as a function to other thermodynamic functions and their derivatives.

Thus, in the case of more complex bodies and thermal processes, where entropy equality does not apply, we must be content with entropy inequality and understand it as a condition imposed by the function called entropy. Although this function itself may remain unknown in its explicit form, even somewhat mystical, it can supply us through this condition with concrete conditions of progress and means of description for thermodynamic processes or the bodies involved in them. We have shown several examples of this in this chapter, including those when the entropy inequality eventually reduces to an equality.

References

1. Thomson, W.: On a universal tendency in nature to the dissipation of mechanical energy. Philos. Mag. **4**, 304–306 (1852)
2. Truesdell, C.: Rational Thermodynamics, 2nd edn. Springer, New York (1984)
3. Owen, D.R.: A First Course in the Mathematical Foundations of Thermodynamics. Springer, New York (1984)

Chapter 13
Multicomponent Bodies and Mixtures

Abstract Independent variables of functions are extended to account for the effects of composition in the case of mixtures; chemical reactions are also included. Entropy inequality is applied on the basic mixture model and chemical reactions are shown to cause non-equilibrium behaviour. The equilibrium in a chemically reacting mixture is analysed. The chemical potentials are introduced and their role as descriptors of compositional effects is stressed. A more general concept of partial quantities is then presented. The analysis of mixtures is developed on the basis of both Helmholtz and Gibbs energies and in both specific and molar quantities. The consequences of reaction stoichiometry are explained, and the extent of the reaction and the reaction Helmholtz and Gibbs energies are introduced. Finally, the basic model of a multiphase body is described, and the criteria of the phase equilibrium are derived.

13.1 Extension of Independent Variables

So far, we have not dealt at all with the chemical composition of our body or system, as it is commonly called in traditional thermodynamics, with the effect of composition on its properties and behaviour. This is because we have so far considered single-component bodies, pure substances or at least uniform bodies of a single and given (unchanging) composition. It is in chemistry that multi-component cases, mixtures, are much more common, where we are also interested in the influence of their composition. Solutions are a typical example. We extend our basic thermodynamics to the case of mixtures easily, in a step that we might call semi-axiomatic. We start from the generalization summarized in Sect. 12.3 and from the fact that to our two basic independent variables, temperature and volume, we simply add other variables describing the content of the individual components, of which there are n in total. Specifically, the masses of the components $m_\alpha,\ \alpha = 1, 2, ..., n$. Thus, we change the functions $\hat{X}(T, V)$ into functions $\tilde{X}(T, V, m_1, m_2, ..., m_n)$. Molar amounts, which are used in chemistry, are easily introduced by substituting molar masses for masses.

We will continue to deal with the uniform body for which the first and second theorems apply as we can see them combined into the reduced entropy

© The Author(s), under exclusive license to Springer Nature Switzerland AG 2024
M. Pekař, *The Essentials of Thermodynamics*, Physical Chemistry in Action,
https://doi.org/10.1007/978-3-031-60321-1_13

inequality (12.3). We shall restrict ourselves next to a body of constant (total) mass (m), into which therefore no substance (mass) flows and from which none flows out. We call such a body a closed system. A more general open system cannot be easily captured by a simple uniform model, here we just refer to more advanced theory, for example, in book [1]. Thus, the following applies

$$m_1 + m_2 + \ldots + m_n = m, \tag{13.1}$$

where m is a given constant (not a variable), and therefore

$$\dot{m}_1 + \dot{m}_2 + \cdots + \dot{m}_n = 0. \tag{13.2}$$

So the basic functions in our mixture look like this, for example:

$$U = \tilde{U}(T, V, m_1, m_2, \ldots, m_n) = \bar{U}(T, V, m_2, \ldots, m_n), \tag{13.3}$$

$$S = \tilde{S}(T, V, m_1, m_2, \ldots, m_n) = \bar{S}(T, V, m_2, \ldots, m_n), \tag{13.4}$$

$$F = \tilde{F}(T, V, m_1, m_2, \ldots, m_n) = \bar{F}(T, V, m_2, \ldots, m_n). \tag{13.5}$$

The second equalities follow from Eq. (13.1), since from it we can express the mass of one component by the others, which we have done here for component 1 (without loss of generality).

Changes in the masses of the components are due (only) to chemical reactions in this closed body. Then they are not arbitrary, but determined by the reactions taking place. The way the masses change over time is given by the equations of chemical kinetics. We will not discuss them explicitly, it is sufficient to know that they somehow determine the time derivatives of the masses:

$$\dot{m}_k = \bar{f}_k(T, V, m_2, \ldots, m_n), \quad k = 2, ..., n. \tag{13.6}$$

These are actually additional constitutive equations in the modelled mixture in addition to Eqs. (13.3)–(13.5).

13.2 Entropy Inequality for the Mixture

Let us put the function (13.5) into the corresponding entropy inequality, i.e., Eq. (12.7). After some minor adjustments, we have

$$-T\Sigma = \left(\frac{\partial \bar{F}}{\partial T} + S\right)\dot{T} + \left(\frac{\partial \bar{F}}{\partial V} + P\right)\dot{V} + \sum_{k=2}^{n} \frac{\partial \bar{F}}{\partial m_k}\dot{m}_k \leq 0. \tag{13.7}$$

Using the already known procedure, we derive from this inequality what we did before—Eqs. (10.11) and (10.4), this time with the function $\bar{F}$ instead of the function $\hat{F}$. We cannot derive further similar conclusions using the given procedure, because the time derivatives of the masses are bound by relations (13.6) and cannot be chosen arbitrarily, which our known procedure requires. This leaves entropy inequality of the form:

$$-T\Sigma = \sum_{k=2}^{n} \frac{\partial \bar{F}}{\partial m_k} \dot{m}_k = \sum_{k=2}^{n} \frac{\partial \bar{F}}{\partial m_k} \bar{f}_k \leq 0, \tag{13.8}$$

which implies that chemical reactions are the source of non-equilibrium processes in this mixture.

13.3 The Equilibrium in the Mixture

Let us now look at the equilibrium in the mixture under discussion. We define the equilibrium processes by means of zero (and thus minimum) value of Σ, thus also of dissipation, cf. Sect. 12.4, in the case of this subsection by zero changes of masses of the components of the mixture, cf. also (13.8):

$$\dot{m}_k = 0, \quad k = 2, ..., n, \tag{13.9}$$

at arbitrary values of T and V. Thus, from the chemical kinetics point of view, equilibrium can be defined by zero reaction rates. However, the zero of the entropy change could also be achieved by (only) zero values of the partial derivatives of the Helmholtz energy in Eq. (13.8). This situation is highly exotic in practice at non-zero values of $\dot{m}_k$, in theory, it corresponds to "frozen" reactions, and such an equilibrium is called weak [2]. We will be satisfied with a strong equilibrium, defined by the conditions (13.9) and even by the zero of those partial derivatives in the equilibrium:

$$\frac{\partial \bar{F}(T, V, m_k^{\mathrm{eq}})}{\partial m_k} = 0, \quad k = 2, ..., n. \tag{13.10}$$

The symbol $\bar{F}(T, V, m_k)$ is a simplified notation for the function $\bar{F}(T, V, m_2, \ldots, m_n)$ and the superscript eq denotes equilibrium. Mathematically, we can otherwise define a strong equilibrium by the regularity of the component masses (in equilibrium): $m > m_k^{\mathrm{eq}} > 0$ and the condition $\dot{m}_k \neq 0 \Rightarrow \partial \bar{F}/\partial m_k \neq 0$. The values of m_k^{eq} are obtained by solving Eqs. (13.9) valid for the equilibrium:

$$\bar{f}_k(T, V, m_k^{\mathrm{eq}}) = 0, \quad k = 2, ..., n; \tag{13.11}$$

while assuming the uniqueness of their solutions.

The equilibrium in a chemically reacting mixture is related to a mixture that does not react chemically at all. Let us demonstrate it. The constitutive Eqs. (13.3)–(13.5) remain unchanged, as does the relation for the total mass (13.1). However, we now do not have the kinetic constitutive Eq. (13.6), so we are not limited in our choice of $\dot{m}_k$ when analysing the entropy inequality. Therefore, in addition to the relations (10.11) and (10.4), we derive the relations $\partial \bar{F}/\partial m_k = 0$; this is precisely the condition of equilibrium in the reacting mixture, but here these nullities are not restricted to equilibrium. In the non-reacting mixture, we thus simplify the constitutive Eqs. (13.3)–(13.5) to the form, e.g., $\bar{F}(T, V)$. Thus, only equilibrium processes are possible in it and it is actually the same body as the model discussed in Sect. 12.3, with the only difference that here the constitutive equations depend on the chosen value of the total mass m.

Equations (13.7) and (13.8) allow us to extend our remark, made in Sect. 12.4, about the direction and equilibrium criteria. Under constant volume and temperature, non-equilibrium processes can take place in a uniform mixture (also) in the direction of the decrease of the Helmholtz energy. They reach equilibrium when the Helmholtz energy is minimum and further unchanged. At constant pressure and temperature, a similar conclusion holds with Gibbs energy.

13.4 Chemical Potential

Let us recap that the properties of mixtures depend on their composition; a change in composition affects the quantities that describe the behaviour of the mixture. The effect of a change in the content of the individual components is captured by the partial derivatives of the relevant quantities with respect to the mass (quantity) of the components. For the sake of consistency with the tradition of classical equilibrium chemical thermodynamics, we will discuss these derivatives using the $\tilde{X}$ functions given a few paragraphs above, not the $\bar{X}$, functions we have used so far. We will call the aforementioned derivative of the Helmholtz energy the (specific) *chemical potential* of a given component:

$$\frac{\partial \tilde{F}}{\partial m_i} = g_i, \quad i = 1, ..., n. \tag{13.12}$$

The unit of such a chemical potential is therefore J/kg. The relationship between the partial derivatives of the two functions discussed, $\tilde{F}$ and $\bar{F}$, i.e., between the chemical potentials, is found by substituting from Eq. (13.1) into relationships (13.5) and performing the derivative of the composite function:

$$\frac{\partial \bar{F}}{\partial m_k} = g_k - g_1, \quad k = 2, ..., n. \tag{13.13}$$

Comparing these relations with the conditions (13.10), we see that in equilibrium the chemical potentials are equal. The equality of the chemical potentials can thus be considered as another criterion for equilibrium—until they are equal, the mixture is not in equilibrium.

The derivations of the Helmholtz energy by temperature or volume are not affected by the presence of different component masses between the independent variables, cf. (13.5), which thus implies

$$\frac{\partial \tilde{F}}{\partial T} = -S, \quad \frac{\partial \tilde{F}}{\partial V} = -P. \tag{13.14}$$

For the time derivative of the Helmholtz energy as a function $\tilde{F}$, we obtain from the first Eq. (13.5) and the definition of the chemical potential another version of the Gibbs equation (this time for a mixture):

$$\dot{F} = -S\dot{T} - P\dot{V} + \sum_{i=1}^{n} g_i \dot{m}_i. \tag{13.15}$$

Using the modified definition of Gibbs energy ($G = F + PV$), we rewrite the obtained Gibbs equation in yet another form (the form for G, or its time derivative):

$$\dot{G} = -S\dot{T} + V\dot{P} + \sum_{i=1}^{n} g_i \dot{m}_i. \tag{13.16}$$

This suggests an alternative way of defining the chemical potential. Here we would have to consider the Gibbs energy as (among other things) a function of pressure. Meanwhile, the pressure in our basic mixture model is treated as the same function as the quantities in (13.3)–(13.5), i.e., $P = \tilde{P}(T, V, m_1, m_2, \ldots, m_n)$. This would imply the existence of an inversion of this function for the volume; the existence of a similar inversion for the uniform body (see discussion of Eq. (10.12)) was based on condition (6.8). There is no reason to believe that this will be different in the basic mixture model under discussion. Let us therefore assume the existence of the inversion and hence the existence of the function

$$G = \breve{G}(T, P, m_1, ..., m_n). \tag{13.17}$$

Then the chemical potential is also defined as

$$g_i = \frac{\partial \breve{G}}{\partial m_i}, \quad i = 1, ..., n. \tag{13.18}$$

In classical (chemical) thermodynamics, the chemical potential is preferably defined just by the Gibbs energy as a function of temperature, pressure and composition.

13.5 Partial Quantities

If the volume of a body changes, the thermodynamic quantities for which volume is the independent variable change accordingly. This is similar for the masses of the components of a mixture. This is sometimes called extensivity or additivity with respect to the volume (mass).[1] Mathematically, this means that, for example, the Helmholtz energy is a homogeneous function of the first degree in the relevant quantities:

$$\tilde{F}(T, \alpha V, \alpha m_1, ..., \alpha m_n) = \alpha \tilde{F}(T, V, m_1, ..., m_n), \tag{13.19}$$

where α is an arbitrary real number. For homogeneous functions, Euler's theorem holds, which in our case of Helmholtz energy is

$$V\frac{\partial \tilde{F}}{\partial V} + \sum_{i=1}^{n} m_i \frac{\partial \tilde{F}}{\partial m_i} = \tilde{F}(T, V, m_1, ..., m_n). \tag{13.20}$$

From this theorem, taking into account (13.14), (13.12) and the definition of the Gibbs energy, it follows:

$$G = \sum_{i=1}^{n} m_i g_i. \tag{13.21}$$

Equation (13.21) shows the relationship between the chemical potentials of the components and the Gibbs energy of the (whole) mixture; in other words, it shows how to calculate the Gibbs energy of a mixture from its composition and chemical potentials. Here, the chemical potential represents one of the so-called *partial quantities* (here more precisely, *partial specific quantities*), which allow one to calculate a property of a mixture from knowledge of the composition and those partial quantities in a similar way as Eq. (13.21) shows for G. Traditionally, partial quantities are defined in terms of functions of temperature, pressure and composition, as follows:

$$y_i = \frac{\partial \check{Y}}{\partial m_i}, \quad i = 1, ..., n. \tag{13.22}$$

We already know that these functions imply the inversion of functions with respect to volume, so it is not surprising that we also encounter partial (specific) volume, defined by the partial derivatives of volume (according to the mass of the relevant component) as a function of temperature, pressure and composition.

By combining the Gibbs equation (13.16) with the time derivative of Eq. (13.21), we derive the so-called *Gibbs-Duhem equation*

[1] A similar extensivity holds for internal energy or entropy, hence, for example, for Helmholtz or Gibbs energy.

$$S\dot{T} - V\dot{P} + \sum_{i=1}^{n} \dot{g}_i m_i = 0. \tag{13.23}$$

This is a constraint on (time) changes in chemical potentials, in general on changes in partial quantities.

13.6 Alternative Analysis by Gibbs Energy

Let us show another case where we base the analysis of the entropy inequality on the Gibbs (i.e., not Helmholtz) energy. For this purpose, let us substitute Eq. (10.53) into Eq. (10.51) and then consider the time derivative of the Helmholtz energy in relation (12.7). Finally, we obtain

$$-T\Sigma = \dot{G} - V\dot{P} + S\dot{T} \leq 0. \tag{13.24}$$

From this equation, we feel that it would be appropriate to use the function (13.17) or its modification $\breve{\bar{G}}(T, P, m_2, ..., m_n)$. Let us then insert the latter into inequality (13.24). After some modifications, we have

$$-T\Sigma = \left(\frac{\partial \breve{\bar{G}}}{\partial T} + S\right)\dot{T} + \left(\frac{\partial \breve{\bar{G}}}{\partial P} - V\right)\dot{P} + \sum_{k=2}^{n} \frac{\partial \breve{\bar{G}}}{\partial m_k}\dot{m}_k \leq 0. \tag{13.25}$$

From this equation, we obtain the related relations (10.16) and (10.17):

$$S = -\frac{\partial \breve{\bar{G}}}{\partial T}, \qquad V = \frac{\partial \breve{\bar{G}}}{\partial P}. \tag{13.26}$$

The entropy inequality then remains in the form

$$-T\Sigma = \sum_{k=2}^{n} \frac{\partial \breve{\bar{G}}}{\partial m_k}\dot{m}_k = \sum_{k=2}^{n} \frac{\partial \breve{\bar{G}}}{\partial m_k}\bar{f}_k \leq 0, \tag{13.27}$$

which is related to (13.8). The latter, however, contains the Helmholtz energy as a function of temperature, volume and composition ($\bar{F}$). If we express the Gibbs energy as a function of the same variables ($\bar{G}$), we get a somewhat different form of the entropy inequality. The relation $F = G - PV$ implies for the derivatives appearing in (13.8):

$$\frac{\partial \bar{F}}{\partial m_k} = \frac{\partial \bar{G}}{\partial m_k} - V\frac{\partial \bar{P}}{\partial m_k}. \tag{13.28}$$

Given Eq. (10.4), we can modify the last term on the right-hand side and finally write

$$\frac{\partial \bar{F}}{\partial m_k} = \frac{\partial \bar{G}}{\partial m_k} + V \frac{\partial}{\partial V} \frac{\partial \bar{F}}{\partial m_k}. \tag{13.29}$$

This result can then be substituted into inequality (13.8). By the various derivations in this section, we wanted to point out that functional dependencies need to be treated with caution in thermodynamics and to keep consistently in mind what the independent variables are, what functions we are deriving, etc.

13.7 Molar Formulations

Chemical thermodynamics usually works with molar quantities instead of specific ones. Conversion is easy using the molar mass of the components (M_i). For example, the traditional *molar chemical potential* with the symbol μ_i, commonly referred to simply as the chemical potential, is obtained directly as follows:

$$\mu_i = M_i g_i, \quad i = 1, ..., n. \tag{13.30}$$

Its unit is J/mol. We can convert functions in a similar way, let's illustrate this with function (13.17) (n_i is the number of moles of component i):

$$G = \breve{G}(T, P, m_1, ..., m_n) = \breve{G}(T, P, n_1 M_1, ..., n_n M_n) = \breve{G}(T, P, n_1, ..., n_n). \tag{13.31}$$

From equalities (13.31), using (13.30) and (13.18), it follows:

$$\frac{\partial \breve{G}}{\partial n_i} = \frac{\partial \breve{G}}{\partial m_i} \frac{\partial m_i}{\partial n_i} = g_i M_i = \mu_i = \frac{\partial \breve{G}}{\partial n_i}. \tag{13.32}$$

The differential of the function $\breve{G}$ of Eq. (13.31) is, in terms of Eq. (13.16),[2]

$$d\breve{G} = -SdT + VdP + \sum_{i=1}^{n} \mu_i dn_i. \tag{13.33}$$

Using the well-known condition (sometimes called Cauchy's condition) for the existence of a total differential in Eq. (13.33), we derive the following relations:

$$\frac{\partial \breve{\mu}_i}{\partial T} = -\frac{\partial \breve{S}}{\partial n_i} \equiv -S_i, \qquad \frac{\partial \breve{\mu}_i}{\partial P} = \frac{\partial \breve{V}}{\partial n_i} \equiv V_i \tag{13.34}$$

[2] Do not forget that, e.g., $\dot{T} = dT/dt$.

where S_i is the partial molar entropy and V_i is the partial molar volume. The relations just derived for the partial molar quantities are analogous to the relations (13.26) for the corresponding mixture quantities. With their help, we can express the total differential of chemical potentials as functions of temperature, pressure and composition, the analogue of (13.33):

$$d\breve{\mu}_i = -S_i dT + V_i dP + \sum_{i=1}^{n} \frac{\partial \breve{\mu}_i}{\partial n_i} dn_i. \tag{13.35}$$

13.8 The Consequences of Reaction Stoichiometry

If the changes in the composition of the mixture indicated by Eq. (13.6) are determined only by chemical reactions, they are related to each other by their stoichiometries. We will show the consequences on a single reaction, because it is easy to illustrate, but on Helmholtz energy, which is not common in traditional chemical thermodynamics. We leave the formulation of the consequences in traditional Gibbs energy to the reader as a fairly simple exercise, especially since the main consequence (the reaction Gibbs energy) is the same, as we will recall at the end of this section. We will formally write the reaction as

$$0 = \sum_{i=1}^{n} \nu_i \mathrm{A}_i, \tag{13.36}$$

where ν_i are the stoichiometric coefficients, negative for the reactants, positive for the products and the symbols A_i represent the chemical formulae of the components of the reacting mixture. The number of molecules reacted and formed in reactions is determined just by the stoichiometric coefficients. Then the quantity defined by the equation

$$\frac{(n_i - n_i^0)}{\nu_i} = \xi, \tag{13.37}$$

where n_i^0 is the initial amount of the i-th component and n_i its amount at some time, is independent of the choice of component and has the same value for all. We call it the *extent of the reaction*. By differentiating the definition of the extent, we obtain another well-known relation

$$d\xi = \frac{dn_i}{\nu_i}. \tag{13.38}$$

Using the reaction extent, we can express the masses of the components

$$m_i = m_i^0 + \xi M_i \nu_i \tag{13.39}$$

and then substitute the independent variables in the function $\bar{F}$:

$$\begin{aligned}\bar{F}(T,V,m_2,\ldots,m_n) &= \bar{F}(T,V,m_2^0+\xi M_2\nu_2,\ldots,m_n^0+\xi M_n\nu_n)\\ &= \bar{\bar{F}}(T,V,\xi).\end{aligned} \tag{13.40}$$

Thus, we have replaced the $n-1$ masses—the independent variables, with a single independent variable—the extent of the reaction. The analogy to the entropy inequality (13.7) is then the inequality

$$-T\Sigma = \left(\frac{\partial\bar{\bar{F}}}{\partial T}+S\right)\dot{T}+\left(\frac{\partial\bar{\bar{F}}}{\partial V}+P\right)\dot{V}+\frac{\partial\bar{\bar{F}}}{\partial\xi}\dot{\xi}\le 0. \tag{13.41}$$

We derive the relations (13.14) in a familiar way, but this time with the function $\bar{\bar{F}}$, and the "residual" entropy inequality—the alternative of (13.8):

$$-T\Sigma = \frac{\partial\bar{\bar{F}}}{\partial\xi}\dot{\xi}\le 0. \tag{13.42}$$

The definition of extent implies:

$$\dot{\xi} = \frac{\dot{m}_k}{M_k\nu_k} = \frac{\bar{f}_k}{M_k\nu_k} \tag{13.43}$$

and after substituting into (13.42)

$$-T\Sigma = \frac{1}{M_k\nu_k}\frac{\partial\bar{\bar{F}}}{\partial\xi}\bar{f}_k\le 0. \tag{13.44}$$

Amount of reactants with a negative stoichiometric coefficient decreases, amount of products with a positive stoichiometric coefficient increases, so that the fraction $\bar{f}_k/\nu_k$ is always positive. Thus, the reaction proceeds in the direction of the negative derivative of $\partial\bar{\bar{F}}/\partial\xi$ or in the direction of the decrease of the Helmholtz energy. Since, in equilibrium, the equality applies in (13.44), the Helmholtz energy is constant and therefore minimal. The reaction therefore proceeds in the direction of minimization of the Helmholtz energy, the corresponding minimum being reached at equilibrium.

For the sake of completeness and emphasis, let us still present the relations between the derivatives according to the range of different types of functional expressions of the Helmholtz energy:

$$\frac{\partial\bar{F}}{\partial\xi} = \sum_{k=2}^{n}\frac{\partial\bar{F}}{\partial m_k}\frac{\partial m_k}{\partial\xi} = \sum_{k=2}^{n}(g_k-g_1)M_k\nu_k = \frac{\partial\bar{\bar{F}}}{\partial\xi} \tag{13.45}$$

(we have used Eq. (13.13)). We will show immediately what these relations can be useful for in the following example. We modify the third expression in Eq. (13.45) as follows:

$$\sum_{k=2}^{n}(g_k - g_1)M_k\nu_k = \sum_{k=2}^{n} g_k M_k \nu_k - g_1 \sum_{k=2}^{n} M_k \nu_k. \tag{13.46}$$

Equation (13.1) actually says that the total mass is still equal to the total initial mass:

$$\sum_{i=1}^{n} m_i = \sum_{i=1}^{n} m_i^0. \tag{13.47}$$

For the sum of the masses, we can substitute from Eq. (13.39):

$$\sum_{i=1}^{n} m_i = \sum_{i=1}^{n} m_i^0 + \xi \sum_{i=1}^{n} M_i \nu_i \tag{13.48}$$

and comparing the last two equations, we obtain

$$\sum_{i=1}^{n} M_i \nu_i = 0 \quad \Rightarrow \quad M_1\nu_1 = -\sum_{k=2}^{n} M_k \nu_k. \tag{13.49}$$

Finally, by substituting the obtained relation into (13.46), given (13.45) and (13.30), we have

$$\sum_{i=1}^{n} g_i M_i \nu_i \equiv \sum_{i=1}^{n} \nu_i \mu_i = \frac{\partial \bar{\bar{F}}}{\partial \xi} = \frac{\partial \bar{F}}{\partial \xi}. \tag{13.50}$$

In accordance with tradition, we will call the expression $\sum_{i=1}^{n} \nu_i \mu_i$ the *Helmholtz reaction energy* and assign it the somewhat unfortunate, but again compatible with tradition, the $\Delta_r F$ symbol. Thus, at the same time, e.g., $\Delta_r F = \partial \bar{\bar{F}} / \partial \xi$. Because of the universality of the chemical potential that we encountered above, for example, around Eq. (13.18), the same sum of molar chemical potentials (weighted by the appropriate stoichiometric coefficients) defines the *Gibbs reaction energy* ($\Delta_r G = \partial \check{\check{G}} / \partial \xi$), much more common in chemical thermodynamics. Here we used the symbol $\check{\check{G}}(T, P, \xi)$.

In this Sect. 13.8, we discussed something about the course of the reaction in the direction of the decrease in Helmholtz energy. In the light of relation (13.50), we can express this criterion for the direction of the reaction using chemical potentials. Roughly, the reaction proceeds in the direction of decreasing chemical potentials. Thus, it proceeds from reactants with higher chemical potentials to products with lower chemical potentials. Of course, by this, we mean the sum of the chemical potentials of the reactants or products weighted by their respective stoichiometric coefficients, as shown in the notation in relation (13.50).

13.9 Multiple Phases

We conclude our macroscopic explanation of the nature of thermodynamics with a very simple example of a non-uniform body model. It will be a non-uniformity somewhat spurious, for it will be a body in the form of a collection of several uniform bodies—*phases*. The phases differ from each other, e.g., in chemical composition, they are separated by boundaries, but they can exchange matter (substances) or energy across them. An example of this would be chunks of ice (solid phase) floating in liquid water, for example in a beaker, or a mixture of a solution of a compound in water, a solid hydrate of that compound and ice (three phases in total). There are not many simultaneously occurring phases in real life, but let us denote the total number of phases by Π. The individual phases will be denoted by the numbers in parentheses: (1), (2),..., (Π). Such a body is made up of a total of n components. The total mass of each of them is a known constant given by the sum of the contributions of each phase:

$$m_i = \sum_{j=1}^{\Pi} m_i^{(j)}, i = 1, 2, \ldots, n. \tag{13.51}$$

The contributions of the phases, $m_i^{(j)}$, are generally not constant (this is the mass exchange allowed).

The internal energy or entropy of the entire multiphase body is given by similar sums, following constitutive equations that also show the chosen independent variables:

$$U = \sum_{j=1}^{\Pi} \tilde{U}^{(j)}(T, V^{(j)}, m_i^{(j)}), \tag{13.52}$$

$$S = \sum_{j=1}^{\Pi} \tilde{S}^{(j)}(T, V^{(j)}, m_i^{(j)}). \tag{13.53}$$

In these we used the shorthand notation $m_i^{(j)} = m_1^{(j)}, m_2^{(j)}, \ldots, m_n^{(j)}$.

We assume (it is part of our simple multiphase body model) the temperature and pressure to be the same in all phases; the pressure is given by the following constitutive equation:

$$P = \tilde{P}(T, V^{(1)}, \ldots, V^{(\Pi)}, m_1^{(1)}, \ldots, m_1^{(\Pi)}, \ldots, m_n^{(1)}, \ldots, m_n^{(\Pi)}). \tag{13.54}$$

If we define the free (Helmholtz) energy of phase j as $F^{(j)} = U^{(j)} - TS^{(j)}$, then, given the constitutive equations above, we have

$$F = \sum_{j=1}^{\Pi} \tilde{F}^{(j)}(T, V^{(j)}, m_i^{(j)}).$$

We insert this definition of the Helmholtz energy of the entire multiphase body into the reduced entropy inequality (12.7):

$$\sum_{j=1}^{\Pi} \frac{\partial \tilde{F}^{(j)}}{\partial T} \dot{T} + \sum_{j=1}^{\Pi} \frac{\partial \tilde{F}^{(j)}}{\partial V^{(j)}} \dot{V}^{(j)} + \sum_{j=1}^{\Pi} \sum_{i=1}^{n} \frac{\partial \tilde{F}^{(j)}}{\partial m_i^{(j)}} \dot{m}_i^{(j)} + P\dot{V} + S\dot{T} \leq 0. \tag{13.55}$$

Equation (13.51) implies that not all masses or their time derivatives are independent. Without loss of generality, we express the masses of the last phase (as dependent):

$$\dot{m}_i^{(\Pi)} = -\sum_{k=1}^{\Pi-1} \dot{m}_i^{(k)}.$$

Next, let us introduce chemical potentials in each phase:

$$g_i^{(j)} = \frac{\partial \tilde{F}^{(j)}}{\partial m_i^{(j)}}.$$

Then we can rewrite the inequality (13.55) in the form:

$$\Big(\sum_{j=1}^{\Pi} \frac{\partial \tilde{F}^{(j)}}{\partial T} + S\Big)\dot{T} + \Big(\sum_{j=1}^{\Pi} \frac{\partial \tilde{F}^{(j)}}{\partial V^{(j)}} + P\Big)\dot{V}^{(j)} + \sum_{k=1}^{\Pi-1} \sum_{i=1}^{n} g_i^{(k)} \dot{m}_i^{(k)} - \sum_{k=1}^{\Pi-1} \sum_{i=1}^{n} g_i^{(\Pi)} \dot{m}_i^{(k)} \leq 0. \tag{13.56}$$

The last two members can be merged:

$$\sum_{k=1}^{\Pi-1} \sum_{i=1}^{n} (g_i^{(k)} - g_i^{(\Pi)}) \dot{m}_i^{(k)}.$$

The entropy inequality must hold for all values of mutually independent and arbitrary time derivatives that occur in it. Thus, by our well-known procedure, we obtain the following equations from it:

$$\sum_{j=1}^{\Pi} \frac{\partial \tilde{F}^{(j)}}{\partial T} = -S,$$

$$\sum_{j=1}^{\Pi} \frac{\partial \tilde{F}^{(j)}}{\partial V^{(j)}} = -P$$

and

$$g_i^{(k)} = g_i^{(\Pi)}, \quad i = 1, \ldots, n; k = 1, \ldots, \Pi - 1. \tag{13.57}$$

We can see from these equations that, in the entropy inequality (13.55) or (13.56), there is always an equation, i.e., the body is in equilibrium, namely in phase equilibrium, or all phases are in equilibrium. The last equation, (13.57), is then a well-known *criterion of phase equilibrium*—the chemical potentials of each component are the same in all phases.

References

1. Pekař, M., Samohýl, I.: The Thermodynamics of Linear Fluids and Fluid Mixtures. Springer, Cham (2014). See also https://www.fch.vut.cz/vav/sapelite
2. Samohýl, I.: Rational Thermodynamics of Chemically Reacting Mixtures. Academia, Praha (1982). In Czech

Chapter 14
The Molecular Model Again

Abstract A microscopic (statistical) view on particle movement is explained. First, the Boltzmann distribution is exemplified by the barometric formula. Then, particle duality is presented and the basic wave equation and its solution are described. An overview of Schrödinger's equation is presented and the essentials of energy levels are shown. The partition function as a key quantity of the discussed molecular approach is defined. Links between the partition function and thermodynamic quantities, internal energy, Helmholtz energy, entropy and others are developed. A molecular-statistical interpretation of heat and work is given.

14.1 Particle and Energy Distributions

We left our molecular (particle) model right after the first opportunity of its use, the kinetic theory of the ideal gas [1]. It is time to revisit it and see if it provides us with some microscopic interpretation of the phenomenological thermodynamic quantities or relationships we have since been concerned with. Just as it has provided us with an interpretation of the pressure of a gas in terms of the momentum transfer of gas particles (molecules) as they collide, or temperature as their mean kinetic energy. We will continue just with pressure again. Although we were dealing with actually macroscopic bodies, we tacitly assumed that they were not so large and we could neglect gravity. But that's not always possible with sufficiently large bodies. For example, the Earth's atmosphere, that is, gas, is known to decrease in pressure with increasing altitude. This is not just a consequence of the simultaneous decrease in temperature, as we will now show in a simple model.

Let us imagine such an atmosphere as a kind of column (of cross-section A) of ideal gas, which, moreover, has the same temperature everywhere. We suspect that the pressure in this column will decrease with its height. This is actually because the gas at a given height must bear the weight of the entire layer of gas above that height in addition to the pressure given by the motion of the gas particles. As the pressure decreases with height, so does the number of gas particles that cause that gravity (and vice versa). Thus, searching for a change in pressure as the height increases

© The Author(s), under exclusive license to Springer Nature Switzerland AG 2024

M. Pekař, *The Essentials of Thermodynamics*, Physical Chemistry in Action,
https://doi.org/10.1007/978-3-031-60321-1_14

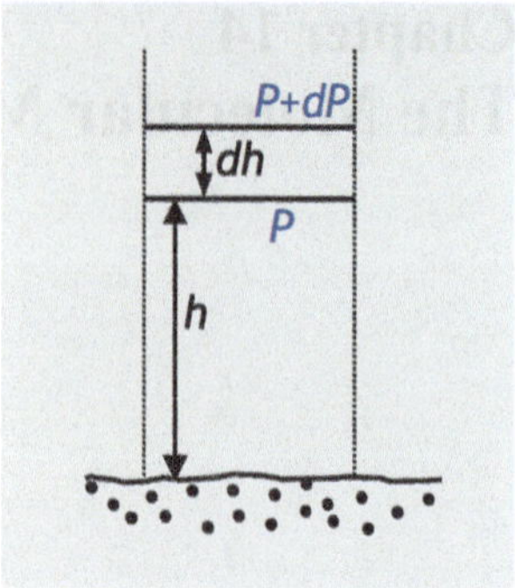

Fig. 14.1 To derive the pressure change in the gas column (atmosphere)

is equivalent to searching for a change in the number of gas particles as the height increases. More precisely, the number of particles per unit volume of gas (denoted N_V), to avoid the dependence of the number of particles on the size of the gas column (see Fig. 14.1). Let us examine the change in pressure as the height changes from h to $h + dh$. The relationship between the pressure of an ideal gas and N_V is easily obtained from its equation of state: $P = N_V kT$, where k is Boltzmann's constant. We then express the change in pressure we are looking for by its differential as a function of N_V: $dP = kTdN_V$. This change in pressure is given by the change in gravity between the heights h and $h + dh$ per unit column cross-section. If all particles are of the same mass m, the change in gravity is due to the amount of particles in the column volume between these heights, i.e., the volume Adh. This change per unit cross-sectional area is therefore equal to $mgN_V Adh/A$. Thus, we have the following equality:

$$\frac{dN_V}{dh} = -\frac{mgN_V}{kT} \tag{14.1}$$

(the negative sign in it is due to the fact that the pressure, and hence the concentration of particles, decreases with height). If we integrate from zero height, where the number of particles per unit volume is N_V^0, to the height h, we get

$$N_V = N_V^0 \exp\left(-\frac{mgh}{kT}\right) = N_V^0 \exp\left(-\frac{E_p}{kT}\right), \tag{14.2}$$

where E_p is the potential energy of the gravitational field (acting on a particle of mass m at height h). The equation obtained describes the concentration distribution of an ideal gas particle in a gravitational field (at a given temperature), the distribution is exponential and its parameter is the potential energy. The higher the potential energy at a given location, the lower the particle concentration there. We can write a very similar equation for pressure because, according to the relation given a few lines above, it is directly proportional to the quantity N_V; it is called the *barometric formula*.

Equation (14.2) is an example of the so-called *Boltzmann distribution*, which we will also encounter in other cases where energy is relevant to the distribution [1]. We encounter distributions in statistics, which deal with large numbers and numerical characteristics of something. This is typical for molecules, of which there are a huge number even in a tiny body, and they do not all have the same energy, for example, but each one has generally different, yet in aggregate distributed only over some interval of energy values. Energy is the main object of interest in thermodynamics—so what does it look like on such a microscopic scale? Not only molecules, atoms and especially the particles of which they are composed (e.g., electrons) but also light behave somewhat strangely compared to macroscopic bodies. In some experiments, they look like these bodies, i.e., tiny material particles (e.g., in photoelectric phenomena, in light we call these particles photons), in other experiments like waves (e.g., in diffraction—bending on a small hole through which they pass). Waves are peculiar in terms of their motion, and hence their associated energy, so we will stop at them briefly.

14.2 Particles and Waves

14.2.1 The Basic Wave Model

Let us recall the description of perhaps the simplest wave, the one that runs through a flexible, "rubber" string when it is deflected and released (and perhaps repeatedly) by some force [2]. We will not give a thorough but simplified derivation and briefly look at the solution of the resulting equation.

Imagine a string spread along a horizontal axis that can only undulate in a direction perpendicular to that axis. Let us denote the deflection of the string during the waves by u, see Fig. 14.2. It will obviously depend on at what point and time we look at the string and its deflection; it will therefore be a function of time and position: $\hat{u}(t, x)$. The change in deflection as we move along the string from location x to location $x + dx$ is described by the corresponding derivative, i.e., $(\partial u/\partial x)dx$. That partial derivative is actually the slope of the string tangent at that location (imagine using Fig. 14.2). As we move along the string, the directive changes, and this is captured by its derivative with respect to the position coordinate: $\partial^2 u/\partial x^2$. This is called the relative elongation.

The string is deflected by the applied force, which induces an elastic force (action and reaction) in it. In the theory of elasticity, Hooke's law is known, which in the case of a simple string states that the elastic force acting per unit volume of the string is directly proportional to the relative elongation: $K\partial^2 u/\partial x^2$. The positive constant of proportionality K is called Young's modulus of elasticity (its unit is Pa). According to Newton's law, the force is equal to the product of mass and acceleration. The

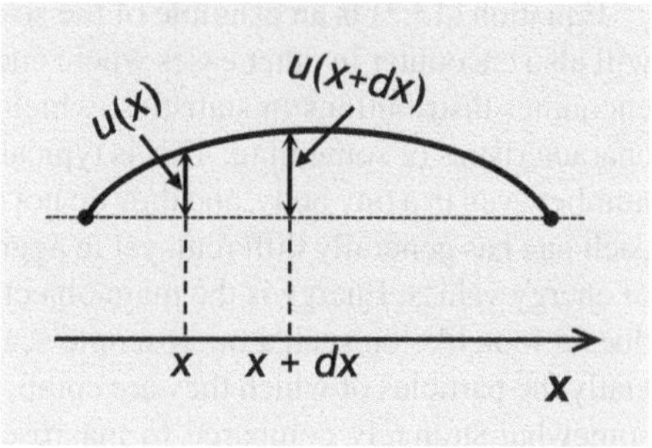

Fig. 14.2 The undulating string and the symbols used in the text

acceleration is given by the second derivative of the trajectory, i.e., of the deflection in the case of our string, with respect to time. We can therefore write

$$\rho \frac{\partial^2 u}{\partial t^2} = K \frac{\partial^2 u}{\partial x^2} \tag{14.3}$$

(ρ is the mass per unit volume, i.e., the density, we consider it constant) and after some modification

$$\frac{\partial^2 u}{\partial x^2} - \frac{1}{v^2} \frac{\partial^2 u}{\partial t^2} = 0, \quad v^2 = K/\rho. \tag{14.4}$$

The quantity v is in units of ms^{-1} and represents the speed of the wave propagating along the string. Equation (14.4) is a (simple) *wave equation*.

Solving the above wave equation is not that complicated, but we will not go through it, the result alone will suffice [3]. By substituting into Eq. (14.4), we can easily find that its solution is the following function:

$$\hat{u}_n(t, x) = (A_n \cos \omega_n t + B_n \sin \omega_n t) \sin \frac{n\pi x}{l}, \quad n = 1, 2, ..., \tag{14.5}$$

where l is the length of the string, A_n and B_n are constants and $\omega_n = n\pi v/l$. So our wave equation has many solutions, in fact infinitely many, numbered by n. The solution, the general solution, is also the (infinite) sum of these functions-solutions. We do not see energy, the quantity we are most interested in here, directly in the wave equation, but we do see at least a piece of kinetic energy (v^2).

14.2.2 The Wave Nature of Particles

The issue of the behaviour of, for example, electrons or photons, here as material particles, here as waves (this is called dual character, duality), was proposed by de

Broglie[1] to be solved by the following idea, or rather postulate. If, for example, light displays this duality, so does matter; let us look for the combination of these two faces in momentum, one of the quantities characterizing motion. Thus, for example, a photon of mass m moving at the speed of light (c) has momentum mc as a particle. Its energy (E) is equal to mc^2 according to the theory of relativity, and so its momentum can be expressed by the relation $p = E/c$. According to quantum theory, the energy of a (light) wave of frequency ν or wavelength λ is equal to $h\nu$ or hc/λ (h is Planck's constant). Thus, by substituting into the expression for the momentum (p), we obtain the relation

$$p = \frac{h}{\lambda}, \tag{14.6}$$

which, according to de Broglie, applies to both the wave and the particle. Then a particle of mass m, moving with velocity v corresponds to (de Broglie's) wavelength $\lambda = h/(mv)$.

14.2.3 Schrödinger Equation

In our basic example of the wave equation, the function $\hat{u}(t, x)$, which was the solution to this equation, described the wave of the string. The wave behaviour of dual particles is similarly described by the wave function obtained by solving the corresponding wave equation, here called the Schrödinger equation. We will just show the basic nature of the path to this equation. First, let's make a little modification, simplifying the above function $\hat{u}_n(t, x)$ as a solution to the wave equation. Using the relations for the sine or cosine of the sum, we find quite easily that the expression $A \cos \omega t + B \sin \omega t$ can be written in the form $\bar{A} \cos(\omega t + \phi)$, where $\bar{A}$ and ϕ are constants combining A and B. The quantity $\bar{A}$ represents the amplitude (maximum deflection) of the wave and ϕ is called the phase angle. The solution of the wave equation can thus be rewritten as [3]

$$\hat{u}_n(t, x) = \bar{A}_n \cos(\omega_n t + \phi_n) \sin \frac{n\pi x}{l}. \tag{14.7}$$

For further interpretation, we will be satisfied with an even more simplified notation in which we collect everything that is not a function of time into a function $\hat{\psi}(x)$ without writing the subscripts n: $\hat{u}(t, x) = \hat{\psi}(x) \cos \omega t$. We have also used the equation $\omega = 2\pi\nu$ and $v = \nu\lambda$. Substituting this simplification into the wave equation (14.4), we have

$$\frac{d^2\hat{\psi}}{dx^2} + \frac{4\pi^2}{\lambda^2}\hat{\psi}(x) = 0. \tag{14.8}$$

---

[1] Louis de Broglie (1892–1987), a French physicist, professor at the University of Paris. He received the Nobel Prize in Physics in 1929 for his discovery of the wave nature of the electron.

Now we finally get the energy into the wave equation, which is what we are most interested in here. In general, the energy of a particle is given by the sum of its kinetic and potential energy (that is, the energy of the force fields acting on or between the particle, such as the electrostatic field). The kinetic energy can be expressed in de Broglie's spirit by the momentum of the particle ($p = mv$); so then:

$$E = \frac{p^2}{2m} + \hat{E}_p(x). \tag{14.9}$$

Using this equation and the de Broglie wavelength ($\lambda = h/p$), we obtain the expression

$$\lambda = \frac{h}{\sqrt{2m[E - \hat{E}_p(x)]}}, \tag{14.10}$$

which we substitute into the modified wave equation (14.8). The result is just the *Schrödinger equation*, or its time-independent variant:

$$\frac{d^2\hat{\psi}}{dx^2} + \frac{2m}{\hbar^2}[E - \hat{E}_p(x)]\hat{\psi}(x) = 0 \tag{14.11}$$

($\hbar = h/2\pi$). We can rewrite it in the form:

$$-\frac{\hbar^2}{2m}\frac{d^2\hat{\psi}}{dx^2} + \hat{E}_p(x)\hat{\psi}(x) \equiv \left[-\frac{\hbar^2}{2m}\frac{d^2}{dx^2} + \hat{E}_p(x)\right]\hat{\psi}(x) = E\hat{\psi}(x), \tag{14.12}$$

which nicely shows that the "action" of a certain derivative and potential energy, i.e., the expression in square brackets, on the function $\hat{\psi}(x)$ is the same as multiplying this function by the energy (a real number). By solving the Schrödinger equation, we obtain the true wave function $\hat{\psi}(x)$, which describes the wave character of the particle(s). More precisely, a stationary wave function (the more general Schrödinger equation containing time is left aside).

14.2.4 Solution of the Schrödinger Equation

To solve the Schrodinger equation, we need to know the specific form of the function $\hat{E}_p(x)$, which depends on the particular problem we want to solve. It is sufficient for us here to say, easily verifiable, that in the simplest case, when there is no potential energy (the case of the so-called free particle; more precisely, it is not in a certain interval of the coordinate x, outside of it, it is infinitely large: $\hat{E}_p(x) = 0$ for x, e.g., between 0 and a), the solution looks like this:

$$\hat{\psi}(x) = A\cos kx + B\sin kx, \tag{14.13}$$

where

$$k = \frac{\sqrt{2mE}}{\hbar}. \tag{14.14}$$

However, the occurrence of the particle is restricted to the positions $0 \leq x \leq a$, implying the following conditions for the wave function: $\psi(0) = \psi(a) = 0$. These then imply $A = 0$ and $ka = n\pi$, $n = 1, 2, ...$ Substituting into Eq. (14.14) gives us interesting information about the energy:

$$E_n = \frac{h^2 n^2}{8ma^2}, \quad n = 1, 2, ... \tag{14.15}$$

Not only do we have the wave function as a solution to the problem (free particle) but also see that the equation is satisfied by a number of (infinitely many) energies, but these can only take on the values given by Eq. (14.15). This equation determines what allowable "quanta" of energy a free particle can have. We say that the energy is *quantized*, and we call the (natural) number n that numbers those possible energies the quantum number. We then often talk about the n-th energy level and about the energy levels or energy levels of a particle.

14.2.5 Energy Levels of Particles

Such a very brief excursion into the wave description of particles, that is, into quantum mechanics, will suffice. Let us just state that we can get an (infinite) range of energies by solving the Schrödinger equation for any particle, i.e., quite generally for atoms or molecules, for example. As a single example, we will give the energies obtained by solving this equation for the hydrogen atom, i.e., actually the *energy levels* of the electron of the hydrogen atom [3]:

$$E_n = -\frac{e^2}{8\pi\varepsilon_0 a_0 n^2}, \quad n = 1, 2, ... \tag{14.16}$$

Here e is the charge of the proton or electron (the elementary charge), ε_0 the permittivity of the vacuum, and a_0 the Bohr radius—this is given by equation $a_0 = \varepsilon_0 h^2/(\pi m_e e^2)$, in which m_e represents the mass of the electron. And for completeness, let us recall that each quantum number n corresponds to a wave function as one of the solutions of the Schrödinger equation (for hydrogen).

Even a very small material body, which we may be able to perceive with our senses, contains a huge number of atoms or molecules. From the wave description, we can see that even each such particle has a huge number of energy levels at its disposal, not to mention all the particles of the body. Thus, in our interpretation, the question arises as to how the energy (and we mean primarily the internal energy) of a body is distributed into the energy levels of its particles at just a given (constant) temperature. Since we already know that the temperature reflects the kinetic energy of the particles, we intuitively feel that at higher temperatures, it will be easier to occupy the higher energy levels.

We have already seen the energy distribution (and at a given temperature) above for the barometric formula, which we mentioned there as an example of the Boltzmann distribution. Without any proof, we adopt this to describe the distribution of energy over the available energy quantum states.[2] The distribution then describes the frequency of occurrence of states of energy E_i, just as the barometric formula described the frequency of occurrence of particles of the appropriate potential (gravitational) energy, both at a given temperature T and in an equilibrium state. The probability of a state i occurring under these conditions is then given simply by the ratio of the relevant exponential of the Boltzmann distribution and their sum over all available energy states:

$$p_i = \frac{\exp(-E_i/kT)}{\sum_i \exp(-E_i/kT)}. \tag{14.17}$$

The denominator of the fraction in Eq. (14.17) is called the *partition function.*

14.3 Partition Function—The Cornerstone

14.3.1 The Partition Function and Internal Energy

We know about internal energy that it is actually the sum of the (kinetic and potential) energies of all the particles that make up a given body. These are divided into levels at constant and equilibrium temperatures according to the Boltzmann distribution. But from the particles' (molecules, atoms) point of view, the energy of a body is not strictly constant even at a given temperature but fluctuates gently around a certain mean value, which is given statistically, the mean value of all available energies of the particles of the body.[3] The fluctuation can be imagined as the result of the constant "communication" of the body with the thermostat, which just maintains the constant temperature of the body. It does not do this with absolute precision, but by means of constant, tiny exchanges of energy by heat, which then result in the desired mean temperature. From a macroscopic point of view, these fluctuations are negligible, even imperceptible. In statistics, we calculate the mean by using the distribution (distribution function) of a given quantity. In the case of energy and the Boltzmann distribution, the mean of the energy is given (we will not use a special symbol for the mean):

[2] For our purposes, it is sufficient to consider the simpler case where each state (level) corresponds to a single energy value.

[3] We have already worked with the mean value—of the velocity, albeit subtly, in the kinetic theory of the ideal gas.

$$E = \frac{\sum_i E_i \exp(-E_i/kT)}{\sum_i \exp(-E_i/kT)}. \tag{14.18}$$

So that is the internal energy of the body in this extended microscopic, statistical, model. It is a function of temperature and volume, as it was in the previous macroscopic concept, and in mixtures also a function of composition. We can express the internal energy a little differently in this micromodel, so that only the partition function emerges, which has a special status in statistical theory [4]:

$$U = E = -\frac{\partial}{\partial(1/kT)} \ln \sum_i \exp(-E_i/kT) \equiv -\frac{\partial \ln \hat{Z}}{\partial(1/kT)} \tag{14.19}$$

or

$$U = E = kT^2 \frac{\partial \ln \hat{Z}}{\partial T}, \tag{14.20}$$

where we denote the partition function as a function of temperature, volume (and number of particles) by $\hat{Z}$.

14.3.2 The Partition Function and Helmholtz Energy

The tight coupling of internal energy with the partition function allows us to more deeply relate the statistical-microscopic model to the macroscopic (phenomenological) thermodynamics we have derived from the simple heating model. Among other things, we found there that the independently variable quantities temperature-volume (and particle number) correspond to Helmholtz energy in the role of thermodynamic potential, These potentials as functions determine other thermodynamic functions. Substituting Eq. (10.11) into the definition of the Helmholtz energy (10.2), we obtain equation

$$F = U + T\frac{\partial \hat{F}}{\partial T}, \tag{14.21}$$

into which we can insert the partition function according to (14.20). We thus obtain the differential equation for the partition function

$$F - T\frac{\partial \hat{F}}{\partial T} = kT^2 \frac{\partial \ln \hat{Z}}{\partial T}, \tag{14.22}$$

the solution of which yields an explicit relation between the Helmholtz energy and the partition function. We will not carry out the solution in detail here, except to note that it can be obtained by the method of variation of constants applied to the ordinary first-order differential equation for the partition function (14.22). In doing so, we make use of the so-called *third law* in determining the integration constant, i.e., the claim that in the limit to zero temperature, the entropy is zero.[4]

The result is the relation we are looking for

$$F = -kT \ln Z, \tag{14.23}$$

which connects the micro-world, i.e., its partition function model, with the macro-world (Helmholtz energy).

14.3.3 The Partition Function and Entropy

The connection with perhaps the most interesting quantity we have derived in the macroscopic description, entropy, is obtained by substituting (14.23) into (10.11):

$$S = k \ln Z + kT \frac{\partial \ln \hat{Z}}{\partial T}. \tag{14.24}$$

We can further modify it for the case defined by Eq. (14.17), from which we see that $\ln Z = -E_i/kT - \ln p_i$, After substituting it into (14.24) and making fairly simple modifications, we have

$$Sp_i = -kp_i \ln p_i - kT \frac{\partial p_i}{\partial T}. \tag{14.25}$$

Performing the summation across all energy states and remembering that $\sum_i p_i = 1$, we easily arrive at the result

$$S = -k \sum_i p_i \ln p_i. \tag{14.26}$$

This is perhaps the most important relation linking the statistical-microscopic model based on quantum energy levels with the entropy derived in the macroscopic view. Its importance also lies in the fact that, in that microscopic view, we also had temperature and volume (and number of particles) as the basic independent variables, i.e., just as in that macroscopic view. In the statistical view, such a model is called a *canonical ensemble* (more precisely, it is the ensemble of all microscopic states of a modelled body with a given temperature, volume and number of particles—we are modelling an equilibrium state). Equation (14.26) is also the starting point for extending the notion of entropy to non-thermodynamic domains, which is certainly not the intention of our tractate, so we will quickly abandon this temptation. However, we point out that this equation suggests what entropy as a function might look like in a particular statistical perspective, as discussed, for example, in Sect. 9.8.

---

[4] We will not need the third law any more here, so we do not pay more attention to it here.

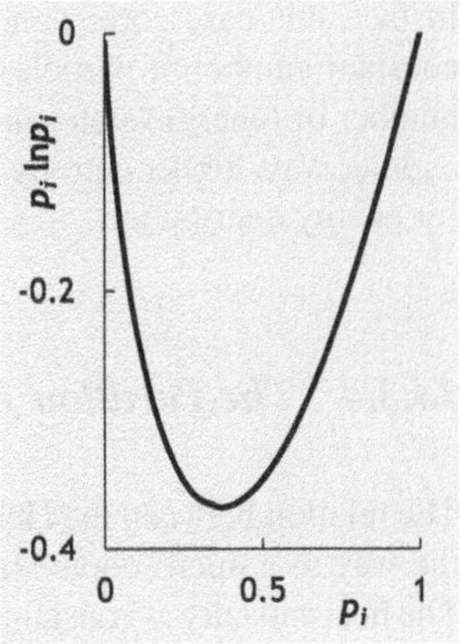

Fig. 14.3 The function $p_i \ln p_i$

Let us pause once more at Eq. (14.26). If only one energy state is possible, then this equation yields a probability equal to one, and then the entropy is zero. This corresponds to a temperature of 0 K and to what we used in solving differential equation (14.22), which led us to Eq. (14.26). At the temperature of absolute zero, substances (compounds, atoms in a crystal lattice, etc.) can really only occupy a single state according to our ideas. Thus, Eq. (14.26) actually calculates entropy relative to its (zero) value at absolute zero.

The individual probabilities (p_i) are somewhere in the interval $(0;\ 1\rangle$. The function $\ln p_i$ is increasing (not only) in this interval. In contrast, the function $p_i \ln p_i$ has a minimum in this interval for a value of (p_i) equal to approximately 0.3679 (exactly equal to 1/e, where e is the Euler number) and is negative throughout the mentioned interval; see Fig. 14.3. Thus, states with probabilities around this value contribute the highest values to the sum in Eq. (14.26) (i.e., to the entropy value), not, for example, states with high probability values. Of course, if states with such probabilities exist in a given body. Similarly, states with low or high probabilities have the smallest values of the $p_i \ln p_i$ function, and thus the smallest contributions to the total entropy. Although the specific values of the probabilities of the possible states also matter, we can roughly say that the more possible energy states, i.e., the more summands in Eq. (14.26), the higher the entropy. Entropy is then explained in a statistical sense as a measure of the number of states available for "storage", the distribution of (total) energy.

Addition at the end. According to (14.17), if all microstates have the same energy, they also have the same probability, $p_1 = p_2 = \ldots = p$. If the number of these microstates is equal to Ω, then since $\sum_{i=1}^{\Omega} p_i \equiv \Omega\, p = 1$, the probability is determined by the number of states: $p = 1/\Omega$. When substituted into (14.26), the famous *Boltzmann relation* for entropy follows:

$$S = k \ln \Omega. \tag{14.27}$$

In fact, this case corresponds to an isolated body, i.e., the adiabatic case (with a constant number of particles or mass). The relation (14.27) simply says that as the number of (energy) states increases, the entropy increases (or vice versa). To put it another way, higher entropy here means the availability of a larger number of states for energy distribution.

14.3.4 The Partition Function and Other Quantities

The relation between the Helmholtz energy and the partition function (14.23) opens the way to a microscopic-statistical expression of other thermodynamic quantities. The heat capacity at constant volume is determined by the Helmholtz energy according to Eq. (10.10). We substitute into it from Eq. (14.23) and see how the partition function actually determines also this capacity:

$$C_V = kT^2 \frac{\partial^2 \ln \hat{Z}}{\partial T^2} + 2kT \frac{\partial \ln \hat{Z}}{\partial T}. \tag{14.28}$$

We have expressed the pressure using the Helmholtz energy by Eq. (10.4); let us add to it from Eq. (14.23):

$$P = kT \frac{\partial \ln \hat{Z}}{\partial V}. \tag{14.29}$$

Knowing the relations between the partition function and the internal or Helmholtz energy, entropy or pressure, we can derive similar relations for the Gibbs energy and enthalpy by substituting in their (possibly modified) definitions ($G = F + PV$, $H = U + PV$):

$$G = -kT \ln Z + VkT \frac{\partial \ln \hat{Z}}{\partial V}. \tag{14.30}$$

a

$$H = kT^2 \frac{\partial \ln \hat{Z}}{\partial T} + VkT \frac{\partial \ln \hat{Z}}{\partial V}. \tag{14.31}$$

However, let's keep in mind that we have introduced functions of temperature and volume (and number of particles), while G or H work as thermodynamic potentials being functions of somewhat different independent variables.

14.4 Statistical Interpretation of Heat and Work

The statistical approach also leads to a molecular interpretation of heat and work [5]. We first express the internal energy in yet another way, by combining Eqs. (14.17), (14.18) and the first equality in (14.20) as $U = \sum_i p_i E_i$. The time derivative of the internal energy is then

$$\dot{U} = \sum_i E_i \dot{p}_i + \sum_i p_i \dot{E}_i. \tag{14.32}$$

Comparing this with the macroscopically derived first theorem, Eq. (7.14), we find that one term on the right-hand side of Eq. (14.32) will represent heating (consequently heat) and one will represent work (volumetric). The first term captures the changes in internal energy due to changes in the probabilities p_i, i.e., changes in the relative occupancy of the quantum states in question. The quantum states themselves (their energies) do not change. According to quantum theory, this would also imply an unchanging volume, and hence this term would represent heating. Let's check this using entropy. It should therefore apply

$$Q_f = \sum_i E_i \dot{p}_i. \tag{14.33}$$

For the time derivative of entropy, we get from Eq. (14.26)

$$\dot{S} = -k \sum_i (\ln p_i + 1) \dot{p}_i \tag{14.34}$$

and after substitution from Eq. (14.17)

$$\dot{S} = -k \sum_i [-(1/kT)E_i - \ln Z + 1] \dot{p}_i. \tag{14.35}$$

Because $\sum_i p_i = 1$, then $\sum_i \dot{p}_i = 0$ and therefore

$$\dot{S} = (1/T) \sum_i E_i \dot{p}_i. \tag{14.36}$$

By combining Eqs. (14.33) and (14.36), we get $\dot{S} = Q_f/T$, the second theorem derived earlier (macroscopically), and so everything appears to be in order. It means, therefore, that the second summand of the right-hand side of Eq. (14.32), $\sum_i p_i \dot{E}_i$, represents (volume) work.

Thus, the transfer of energy by heat is interpreted by statistical theory as a change in the relative frequency of occupancy of given quantum states; the energy absorbed by heat leads to an increase in the relative occupancy of higher energy states (levels). Energy transfer by work, on the other hand, is explained as changes in the energy

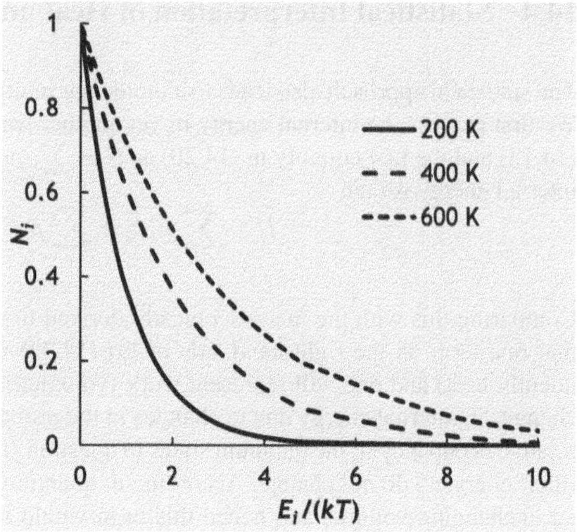

Fig. 14.4 Sketch of the increase in occupancy of higher energy levels with increasing temperature

values of quantum states, while their relative occupancy remains unaffected. We can further extend this interpretation with respect to our simple heating model. The part of the heating not related to the volume change is attributed to the temperature change, which in turn is closely related to the (kinetic) energy of the particles. Increasing the temperature by absorbing energy by heat means increasing the occupancy of higher energy levels. This is quite logical—increased temperature as a consequence of increased energy is associated with the "storage" of the increment of energy in higher levels. This is illustrated in Fig. 14.4, which shows the dependence of the quantity $N_i = \exp(-E_i/kT)$, cf. also Eq. (14.17), on energy (expressed in units of kT) for three different temperatures.

References

1. Widom, B.: Statistical mechanics. A concise introduction for chemists. Cambridge: Cambridge University Press (2002)
2. Kvasnica, J.: Mathematics in physics. Praha: Academia (1989). In Czech
3. McQuarrie, D.A., Simon, J.D.: Physical chemistry: A molecular approach. Sausalito: University Science Books (1997)
4. Malijevský, A.: Lecture notes in statistical thermodynamics. Praha: Vydavatelství VŠCHT (1998). In Czech
5. Denbigh, K.: The principles of chemical equilibrium. Cambridge: Cambridge University Press (1961)

Chapter 15
Conclusion

Let us summarize our journey to the essence of thermodynamics. The foundation, which is often the content of introductory courses in, for example, physical chemistry, can be built quite well on the simple *model of heating* presented in Sect. 6.2. From it, the two most important quantities—*internal energy* (Sect. 7.6) and *entropy* (Sect. 9.1 onwards)—can be derived quite easily in mathematical terms. This basis deals with *uniform bodies* and the whole can be summarized in a few *axioms*, see Sect. 10.6.

The uniform body basis also contains the possibility of the occurrence of only so-called *equilibrium states* or processes, explained in Sect. 12.4. By breaking away from this constraint, we obtain an *entropy inequality*, Chap. 11, in contrast to the previously derived only equality (i.e., a kind of definition of entropy). This allows us to adapt the axiomatic construction of the foundations of thermodynamics to the generalized first two theorems as given in Sect. 12.2, where they were also combined into the so-called reduced entropy inequality as a fundamental principle that must be satisfied.

Again, this modified axiomatization implies that a uniform body of the usual basic interpretation, with temperature and volume as independent variables, is capable of only *equilibrium processes*. This is shown in Sect. 12.3. Section 12.5 illustrates that non-equilibrium processes are possible even in a uniform body, but with a more complex structure of independent variables, or, as follows from Sect. 13.2, if chemical reactions take place in it. The latter section belongs to the chapter on multicomponent uniform bodies (mixtures). That is, to Chap. 13, which thus describes the procedure in the case of multicomponent mixtures.

The concept of a uniform body implies spatial invariance of its properties. The quantities that describe this body are thus not functions of spatial coordinates. They are functions only of time, and even then often only indirectly as a function of (for example and in particular) volume and temperature, which are independent variables that change over time. The description of spatial dependencies is the subject of more advanced thermodynamics, see for example, the book [1], to which this small treatise is a suitable introduction.

© The Author(s), under exclusive license to Springer Nature Switzerland AG 2024
M. Pekař, *The Essentials of Thermodynamics*, Physical Chemistry in Action,
https://doi.org/10.1007/978-3-031-60321-1_15

The phenomenological, macroscopic point of view prevails in the interpretation. Occasional excursions into kinetic or statistical approaches are intended to illuminate the nature of the connection to microscopic theories. Above all, to point out the quantities that bring about that connection, which provide the main macro- and microscopic interpretation.

Somewhat subtly, we have touched on the opening quote in various places, trying to remind the reader that the nature of thermodynamics can also be viewed through the lens of the science of *motion*, using quantities that people have invented and created for this purpose. If dynamics denotes the science of forces, and force is the cause of motion, thermodynamics can be roughly understood as the science of thermal motion and its contexts. So, we touched on the motion of the fundamental particles of bodies (the items in that quote), say molecules, and we have lingered longer on the motion observable to the eye, the macroscopic motion. At this level, whole bodies or parts of bodies move, but of course being made up of molecules.

Reference

1. Pekař, M., Samohýl, I.: The Thermodynamics of Linear Fluids and Fluid Mixtures. Springer, Cham (2014). See also https://www.fch.vut.cz/vav/sapelite

Index

© The Editor(s) (if applicable) and The Author(s), under exclusive license to Springer Nature Switzerland AG 2024

M. Pekař, *The Essentials of Thermodynamics*, Physical Chemistry in Action,
https://doi.org/10.1007/978-3-031-60321-1

SPRINGER NATURE

GPSR Compliance

The European Union's (EU) General Product Safety Regulation (GPSR) is a set of rules that requires consumer products to be safe and our obligations to ensure this.

If you have any concerns about our products, you can contact us on ProductSafety@springernature.com

In case Publisher is established outside the EU, the EU authorized representative is:

Springer Nature Customer Service Center GmbH
Europaplatz 3
69115 Heidelberg, Germany

The manufacturer's authorised representative in the EU is Springer Nature Customer Service Centre GmbH, Europaplatz 3, 69115 Heidelberg, Germany. If you have any concerns regarding our products, please contact ProductSafety@springernature.com

Printed and bound by CPI Group (UK) Ltd, Croydon, CR0 4YY
01/10/2026
02225916-0017